The Acid Rain Debate

Contemporary Issues in Natural Resources and Environmental Policy

Series sponsored by
the Utah State University Ecology Center
and College of Natural Resources

FREDERIC H. WAGNER, *Series Editor*

#1   PREDATOR CONTROL AND THE SHEEP INDUSTRY:
The Role of Science in Policy Formation
Frederic H. Wagner

#2   SAVING THE HIDDEN TREASURE:
The Evolution of Groundwater Policy
Henry C. Kenski

#3   THE ACID RAIN DEBATE:
Science and Special Interest in Policy Formation
Bruce A. Forster

Natural Resources and environmental policy decisions are made for a variety of administrative, economic, and political reasons, with scientific understanding factored in to varying degrees. Policy analysts are increasingly developing a body of theory which generalizes the manner and extent to which scientific data influence policy formation. The purpose of this series is to summarize the key technical information on a variety of important, often controversial, natural resources and environmental issues, and to interface it with the policy considerations. The ultimate goal is to inform policymakers and the public about the scientific information and policy considerations on such issues; and to contribute to that aspect of policy analysis which explores the role of factual information in policy formation. Authors have been selected who are especially knowledgeable on the topics covered. They strive to represent the topics objectively and avoid advocacy for any one decision option, but they do not hesitate to point out incongruences between policy action and the technical information.

The Acid Rain Debate

Science and Special Interests in Policy Formation

Bruce A. Forster

Iowa State University Press / Ames

Bruce A. Forster is dean of the College of Business, University of Wyoming, Laramie.

© 1993 Iowa State University Press, Ames, Iowa 50010
All rights reserved

Authorization to photocopy items for internal or personal use, or the internal or personal use of specific clients, is granted by Iowa State University Press, provided that the base fee of $.10 per copy is paid directly to the Copyright Clearance Center, 27 Congress Street, Salem, MA 01970. For those organizations that have been granted a photocopy license by CCC, a separate system of payments has been arranged. The fee code for users of the Transactional Reporting Service is 0-8138-1684-X/93 $.10.

♾ Printed on acid-free paper in the United States of America

First edition, 1993

Library of Congress Cataloging-in-Publication Data

Forster, Bruce A. (Bruce Alexander)
    The acid rain debate : science and special interests in policy formation / Bruce A. Forster.—1st ed.
       p.   cm.—(Contemporary issues in natural resources and environmental policy ; #3)
    Includes bibliographical references and index.
    ISBN 0-8138-1684-X (alk. paper)
    1. Acid rain—Environmental aspects—United States. 2. Acid rain—Government policy—United States. I. Title. II. Series.
TD195.5.F67   1993
3b3.73'86—dc20   92-47434

Contents

Acknowledgments, vii
About the Author, viii
Editor's Preface, ix

1. **Introduction, 3**
   American-Canadian Research Efforts, 3
   Official Recognition of Transboundary Pollution by the United States, 4
   Official Recognition Reversed, 5
   Executive Order 12,291 and the Political Philosophy of the
       Reagan Administration, 7
   Special Interest Behavior and Policy Formation, 8
   References Cited, 10

2. **Sources of Acid Rain, 12**
   Introduction, 12
   SO_2, NO_x, and Acid Rain, 14
   Transport of Airborne Pollutants: Long Range or Short Range? 20
   The Linearity Hypothesis, 25
   Indicators of Acid Rain, 26
   Time Trends in Rainfall Acidity, 27
   References Cited, 27

3. **Agricultural Impacts, 30**
   Physical Impacts, 30
   Economic Impacts, 39
   References Cited, 44

4. **Forest Impacts, 48**
   Physical Impacts, 48
   Economic Impacts, 54
   References Cited, 55

5. **Aquatic Impacts, 57**
   Physical Impacts, 57
   Economic Impacts, 64
   References Cited, 74

6. **Materials Damage, 78**
   Physical Impacts, 78
   Economic Impacts, 84
   References Cited, 87

7. **Human Health and Visibility Impacts, 90**
   Health Impacts, 90
   Visibility Impacts, 94
   Economic Valuation of Health and Visibility Impacts, 94
   References Cited, 98

8. **Control Options and Strategies, 100**
   Options for Reducing Emissions from the Electric-Power Industry, 100
   Strategies for Implementing Emissions Reductions, 105
   Comparisons of Market versus Command-and-Control Policies, 111
   References Cited, 113

9. **Exaggeration of Control Costs and Disruption, 116**
   Introduction, 116
   Electric Utilities, 118
   Coal Suppliers, 123
   Coal Miners, 126
   Electricity Consumers, 128
   Regional Political Representatives, 129
   References Cited, 132

10. **Scientific Validity and Political Sensitivity, 134**
    Introduction, 134
    The Bureaucratic Game, 134
    References Cited, 143

11. **Non-Convexities, Irreversibilities, and Acid Rain Controls as Insurance, 144**
    Benefit-Cost Estimates and Insurance, 147
    References Cited, 150

Epilogue, 151

Selected General References, 157
Index, 161

Acknowledgments

Reviewer for this volume is Dr. Jack A. Donnan, Senior Economist, Policy and Planning Branch, Ontario Ministry of the Environment. The series editor and I thank him for his helpful comments and the draft manuscript. Thanks also go to the series editor, Dr. Frederic H. Wagner, for his careful editing, constructive suggestions, and encouragement.

This study owes much to the various researchers on acid rain issues cited in this volume. I have had the advantage of interacting with many of these researchers at numerous conferences, peer review sessions, and seminars over the past decade. They have significantly contributed to my understanding of this multi-disciplinary problem and I thank them. Special thanks go to my colleague on several jointly authored papers on acid rain topics, Thomas Crocker, and to Lance Evans, Orie Loucks, Jack Ketchesen, and Douglas Ormrod for improving my knowledge on specific scientific issues.

For her skillful editing and patience, I wish to thank Lynne Bishop. And for her artwork, I thank Linda Marston.

The various drafts have been more than ably typed by Janet Klinker. I thank her for her patience.

My wife Dale and my children are thanked for their patience while the manuscript was being written. A central concern in the acid rain debate is for the future of the environment versus current costs. Thus, I dedicate this book to my son Jeremy and my daughters Kelli and Jessica: their generation will inherit the result of the current governing generation's decision.

About the Author

Bruce A. Forster, born in Toronto, Ontario, received his B.A. in mathematics and economics from the University of Guelph, Canada, and his Ph.D. in economics from the Australian National University. He was a member of the Economics Department of the University of Guelph from 1973 to 1987. In 1987, he joined the faculty of the University of Wyoming, Laramie, as a professor of economics. Forster became dean of the College of Business in 1991. He has been a visiting professor at the University of British Columbia; the University of Newcastle, N.S.W., Australia; the University of Wyoming; and the Academy of International Economic Affairs, Ministry of Economic Affairs, Taiwan (Republic of China).

His research interests have centered on the mathematical modelling of environment-economy interactions in general and the acid rain issue in particular. He has served as a consultant on acid rain economics for the United States Environmental Protection Agency, the Ontario Ministry of the Environment, Environment Canada, Agriculture Canada, the Canada Department of Fisheries and Oceans, and Work Group I of the Joint U.S.-Canada Memorandum of Intent concerning Transboundary Air Pollutants. He was associate editor of the Journal of Environmental Economics and Management from 1989 to 1991.

Editor's Preface

The phenomenon of acid precipitation appears to have begun with the onset of the Industrial Revolution, and was first described by Robert Angus Smith in 1852 in England. In his 1872 book, *Air and Rain,* he linked coal burning in Britain to acid rain. By the turn of the century, fish populations were declining in Norwegian lakes, although the cause was not known at the time. It was not until the 1950s and 1960s that the problem was fully researched and reported in Norway and Sweden. Since then, extensive forest death has been thoroughly investigated in Germany, Poland, and Czechoslovakia.

Research began in North America in the late 1960s and early 1970s. By the 1980s there was extensive U.S. and Canadian scientific literature on the sources and effects of ozone and acid precipitation, both formed as secondary pollutants from nitrogen oxide and in the latter case from sulphur dioxide. Documented effects include damage to agriculture (in the case of ozone); forests; lakes and streams; such human artifacts as metals, masonry, and paint; and human health. The most seriously affected areas are southeastern Canada and northeastern United States.

Despite this extensive understanding of the problem, and decisions by Canada and 18 European countries in the early 1980s to reduce sulphur dioxide emissions by 30%, the United States did not have a formal policy in place until late 1990, and by early 1992 still had no extensive action program. Policy movement started, stopped, and started with changing administrations in Washington: moved forward at the end of the Carter administration, halted during 8 years of the Reagan administration on the grounds of insufficient information, and then moved forward in July 1989 when President Bush proposed amendments to the Clean Air Act. At this point Congress seized the initiative, with both the House and Senate submitting bills in early 1990, but only after considerable internal contention in each case. A compromise bill was finally passed in late 1990, and signed by President Bush on November 15.

In this book, Professor Forster provides lucid insights into the complex of social and economic forces and their politicization that have been responsible for this hesitancy in American action. He reviews the accrued scientific information, which in total has enlightened policy process on the nature and magnitude of the problems but which at times has been shaded or used selectively to support policy positions. And he closes by giving evidence that some of the ecological damage can be economically irreversible, using this to argue for the importance of taking immediate action rather than waiting for more confirmatory evidence.

The entire treatise is a classical case study of the political complexities surrounding the solution of major environmental problems, and the uses and misuses of scientific research in the policy-making process. With professional experience on acid deposition both in Canada and the United States, Professor Forster is especially well qualified to provide a balanced, binational perspective on the issue.

The Acid Rain Debate

CHAPTER 1

Introduction

American-Canadian Research Efforts

The issue of acid rain has attracted widespread attention in North America since the late 1970s. Despite a "collaborative" effort between the United States and Canada that spanned a decade, the two governments were not able to achieve a joint agreement to reduce the emissions of precursors to acid rain during the 1980s.

In October 1978, the Canadian and U.S. governments established the Bilateral Research Consultation Group on the Long Range Transport of Air Pollutants "in response to mutual interest and concerns." This group submitted its first report in October 1979 and a second report in November 1980. These reports were the first comprehensive statements concerning the knowledge of acid rain and its effects (or possible effects) on various receptor categories in eastern North America. In August 1980, the two countries signed a Memorandum of Intent on Transboundary Air Pollution. Under the memorandum, four work groups were established to study various aspects of the acid rain phenomenon (as well as other air pollutants). These work groups essentially superceded the joint Research Consultation Group, which was not formally dissolved. Problems in completing the Report of Work Group 1 on Impact Assessment resulted in the final (Phase 3) report being delayed by 1 year. This report appeared in February 1983.

Phase 4 was to have considered negotiating an agreement to reduce emissions of the precursors of transboundary air pollutants. Phase 4 was quickly superceded by high-level discussions between then Minister for External Affairs Allan MacEachen and Secretary of State George Shultz. In March 1985 following the first "Shamrock" summit, Prime Minister Mulroney and President Reagan appointed two special envoys, William Davis and Drew Lewis, to study the acid rain problem. In the fall 1985 it was expected that the report would recommend a "modest" controls program that would bring a 3- to 4-million-ton reduction in sulfur dioxide (SO_2) emissions. When Lewis and Davis presented their report (February 1986), it did not include any controls program targets for emissions reductions. While concluding that acid rain was a problem, the report called for a "commercial technology demonstration program" at a cost of $5 billion, which would be split between the U.S. government and industry. Davis argued that this was the most Canadians could hope that the United States would accept (Long 1986).

Acid rain continued to be a significant issue at subsequent "Shamrock" summits. There is still no joint agreement. The Canadian government decided to go it alone in 1984 and has since introduced a controls program designed to reduce Canadian emissions by 50%. The United States (along with Great Britain and Poland) declined to participate in the so-called "30% Club" — a group consisting of 18 European nations and Canada, which aim to cut their emissions of SO_2 by over 30% over a 13-year period. In late 1988 the United States did sign an international accord concerning the control of nitrogen oxides that also contribute to acid rain but continued to refuse to join the 30% Club (*Coal Week* 1988).

Why have these two North American countries responded so differently to acid rain controls over this time period despite extensive collaborative efforts?

Official Recognition of Transboundary Pollution by the United States

For most of the 1980s, the United States took the position that there were too many scientific uncertainties concerning the benefits that would be obtained through a very costly controls program. This was not the official view prior to 1981. In April 1980, a letter from the Canadian and American sections of the International Joint Commission to Sharon Ahmad, Deputy Assistant Secretary of State for European Affairs (including Canada), indicated that they believed acid rain was one of the most serious environmental issues facing the two countries (Carroll 1985). The commission recommended vigorous domestic initiatives in both countries.

In the final days of the Carter administration, Douglas Costle, administrator of EPA, announced that emissions from the United States were contributing significantly to the acid inputs over sensitive areas in Canada, and that emissions from both sides of the border were causing significant problems in each country (Wetstone 1984). With the December 1980 passage of the Canadian Clean Air Act, Section 21.1, which provides "reciprocity" in dealing with transboundary pollution, Costle prepared to invoke Section 115 of the United States Clean Air Act. Under Section 115 the EPA administrator may require each state to revise its State Implementation Plan (SIP) to reduce impacts on foreign countries. As Costle was leaving office at the end of 1980, the EPA staff were directed to develop for the next administrator the necessary information concerning which states would be notified. Costle's findings were relayed to then Secretary of State Edmund Muskie in a letter dated January 13, 1981.

Official Recognition Reversed

The process that Costle set in motion was not continued by the next administrator, Anne Gorsuch, who was appointed by the newly elected President Ronald Reagan. Not only did Ms. Gorsuch not issue the SIP revision notices, she stated that Costle's findings were insufficient to invoke Section 115 (Garland 1988). Subsequent administrators William Ruckelshaus and Lee Thomas failed to invoke Section 115 as well.

This inaction by the EPA administrators resulted in a suit being filed against Lee Thomas in his capacity as the administrator of EPA by the state of New York in conjunction with five other states, four environmental organizations, and four individuals (Garland 1988). Federal District Court Judge Norma Holloway Johnson ruled in favor of the plaintiffs. The court ordered a re-evaluation of the reciprocity condition. If it was determined that reciprocity existed, then the EPA was to issue the relevant SIP notices within 180 days (Garland 1988).

Thomas determined that reciprocity did in fact exist but the court order concerning the issuance of SIP notices was stayed while EPA appealed the court's decision (Garland 1988).

The court of appeals overturned the lower court decision, not on the merits of the case concerning acid deposition and its impacts under Section 115, but on procedural grounds. The court argued that Costle's findings were issued without "notice and comment rulemaking" as required by the Administrative Procedures Act (Ohline 1987; Garland 1988).

Ohline (1987) claims that the court of appeals's decision shows the reluctance of the American judicial system to deal with cases involving scientific, technical, economic, and political aspects. According to Ohline (1987) the court could have decided to allow EPA to invoke rule-making procedures at the time it issued the SIP notices to the individual states rather than overturning the lower court decision.

While Ohline (1987) believes that case law supports the court's decision, Garland (1988) disagrees. In fact, Garland (1988) argues that the court's decision was in error because it failed to examine Section 307 of the Clean Air Act, which deals with the relevant administrative procedures to be invoked in executing the Clean Air Act. The relevant subsection, 307(d), does not list Section 115 actions in the group of actions specified as requiring "notice and comment rulemaking." According to Garland (1988), the administrative procedures specified in Section 307 have been used historically for the Clean Air Act in lieu of those of the Administrative Procedures Act, which the court chose to invoke! Garland (1988) also argues that the example in case law that the court chose to rely on was clearly distinguishable from the acid rain case and should not have been used as the basis for a decision. Overall, in assessing the court of appeals's decision, Garland states that its decision ". . . is incongruous with congressional intent."

Both Ohline (1987) and Garland (1988) suggest that political concerns clouded the court's decision. Ohline notes that the Clean Air Act and Costle's findings belonged to previous administrations and that the policy being considered by the court was ". . . contrary to that of the present administration."

In late 1988, the Sierra Club and the Izaak Walton League filed another suit against EPA on this issue with the U.S. Court of Appeals (*Coal Outlook* 1988).

Executive Order 12,291 and the Political Philosophy of the Reagan Administration

Within a month of taking office, President Reagan signed Executive Order 12,291, which requires proposed "major" regulations to have a Regulatory Impact Analysis including a cost-benefit analysis. The following specific requirements under Executive Order 12,291 have particular significance for the acid rain issue (Alviani 1980–1981, 294):

1. Administrative decisions shall be based on adequate information concerning the need for and consequences of proposed government action;
2. regulatory action shall not be undertaken *unless the potential benefits to society from the regulation outweigh the potential costs to society;*
3. agencies set regulatory priorities with the aim of *maximizing the aggregate net benefits to society, taking into account the condition of the particular industries affected by the regulations, the condition of the national economy, and other regulatory actions contemplated for the future* [emphasis added].

The effect of 12,291 is to shift the burden of proof to those who desire increased government action or regulation and to increase the weight to be given to opposing arguments consistent with the president's own economic and political philosophy of government intervention and regulation (Alviani 1980–1981, 308).

The Regulatory Impact Analyses were made subject to review by the Office of Management and Budget. In early 1981, the head of the Office of Management and Budget, David Stockman, was quoted as saying,

> I kept reading these stories that there are 170 lakes dead in New York . . . well how much are the fish worth in these 170 lakes that account for four percent of the lake area of New York? And does it make sense to spend billions of dollars controlling emissions from sources in Ohio and elsewhere if you're talking about a very marginal volume of dollar value . . .? (Norton 1982, 60)

Special Interest Behavior and Policy Formation

Conventional economic theory presumes that individuals (be they consumers, producers, workers, resource owners, etc.) conduct their affairs so as to make themselves as well-off (in terms of happiness, profits, or income) as possible *given their available resources, information, and institutional arrangements or constraints.* In this conventional view, regulations are set by altruistic governments in the public interest in order to correct for perceived failures of the market system. In this antidotal view of regulation, economic agents passively adjust to new regulations and a pareto optimum (socially ideal) outcome is achieved.

Real-world economic agents do not behave so passively when it comes to the introduction of government regulations and governments, and rather than be altruistic they respond to political pressures. The various adjustments in behavior necessitated directly and indirectly by imposing regulations will cause wealth (monetary or psychic) to be transferred between economic agents. These potential wealth transfers create incentives for economic agents to try to affect the formation of government regulation. Hartle (1983) suggests that the government decision-making process can be best understood by considering a set of intersecting "games." The developments or lack of development in the area of acid rain control since 1981 may be best understood by considering the following set of intersecting games: the special-interest group game; the political game; the bureaucratic game. The term "rent-seeking behavior," refers to the activities and the allocation of resources undertaken by individuals (or groups of individuals) in order to achieve changes in government regulations that will transfer wealth to them or to block changes that will transfer wealth away from them. In Hartle's games, *all players* maximize their own utility, playing by the rules of their own games.

In light of the position stated in Executive Order 12,291, it was fairly clear what was required by those who oppose legislation to reduce emissions—they should emphasize scientific uncertainties associated with emissions, deposition and subsequent impacts (i.e., benefits of reducing emissions), while simultaneously decrying the high costs of control and the economic disruption that would result from such controls.

The arguments put forward by the various groups usually contain both elements. Cast in this fashion, the arguments generally do

1: Introduction

not sound like pure self-interest on the surface. A typical comment is that of Mallick (1987), who states, "With all of the uncertainty surrounding acid rain, the one certain consequence is a staggering cost — perhaps exceeding $100 billion over 20 years." However, the incentive exists for various impacted groups to form informal coalitions to lobby against acid rain controls and to exaggerate the costs and disruption that would result. The largest lobby in the United States for 1986–1987 was the Citizens for Sensible Control of Acid Rain (CSCAR) (*Coal Outlook* 1987). CSCAR has raised $4.8 million from industry groups to lobby against acid rain legislation and it has been alleged that CSCAR is disseminating inaccurate information — in particular, exaggerating the increase in electricity costs that will result (*Environmental Action* 1987).

The lay-public frequently expects that natural science produces hard facts that are non-disputable. Unfortunately, this view is as naive as the conventional classroom depiction of policy formation. Science is a process of discovering knowledge. The acid rain phenomenon confronted scientists in a number of disciplines with a new problem that in many cases required new scientific analysis techniques.

This raises the possibility of debate over the appropriate experimental design, as well as debate over the interpretation of results. People frequently think that the "facts will speak for themselves." Unfortunately, observers speak for the "facts" as a result of their interpretation of observations. Individual observers may have incentives to interpret results in a manner that is to their own benefit. This does not have to be the result of dishonest practices. Psychologists use the term "cognitive dissonance" to describe the tendency for individuals to screen out information that does not confirm their beliefs and/or prejudices rather than adjusting their beliefs.

The purpose of this book is to highlight the difficulties encountered in moving from the recognition of a potential policy problem to the implementation of a policy decision, based upon natural- and social-science research, in the specific case of acid rain. While the book seeks to portray our knowledge, the emphasis is on the debate and controversy that has emerged concerning various aspects of the issue. This book shows the difficulty that real-world policymakers face in trying to make decisions when faced with legitimate uncertainty (natural-science and economic), as well as incentives for individuals to interpret analyses to their own advantage.

The debate concerning the sources of acid rain is discussed in Chapter 2. Chapters 3–7 present overviews of the knowledge and debate concerning the physical and economic impacts of acid rain on the various "receptor" categories. Chapter 8 deals with alternative approaches to reducing acid rain. Chapter 9 addresses the incentives that various special-interest groups have to exaggerate the costs of controlling acid rain. The political manipulation of science is considered in Chapter 10. Chapter 11 identifies qualitative features of the impact of acid rain, which suggest that an acid rain controls program should be implemented despite the uncertainties involved in the debate. The Epilogue discusses the changes that took place in the acid rain debate once President Bush assumed office in early 1989.

References Cited

Alviani, J.D. 1980–1981. Federal regulation: The new regimen. Boston Coll. Environ. Affairs Law Rev. 9:285–95.

Carroll, J.E. 1985. Transboundary air pollution: The international experience. In Acid Deposition: Environmental, Economic and Policy Issues. Ed. D.D. Adams and W.P. Page, pp. 507–21. New York: Plenum Press.

Coal Outlook. 1987 October 12.

———. 1988 November 14.

Coal Week. 1988 November 7.

Environmental Action. 1987. "In Sheep's Clothing." September/October: 9.

Garland, C. 1988. Acid rain over the United States and Canada: The D.C. circuit fails to provide shelter under Section 115 of the Clean Air Act while state action provides a temporary umbrella. Boston Coll. Environ. Affairs Law Rev. 16:1–37.

Hartle, D.G. 1983. The theory of "rent-seeking": Some reflections. Can. J. Econ. 16:539–54.

Lewis, D. and W. Davis. 1986. Joint report of the special envoys on acid rain.

Long, C. 1986. Washington report. J. Air Pollut. Control Assoc. 36:118.

Mallick, E. 1987. Report on the 15th APCA government affairs seminar: Quest for clean air–challenges to the 100th Congress. J. Air Pollut. Control Assoc. 37:698–99.

Norton, K. 1982. Symposium keynote address I. In Acid Rain. Ed. P.G. Gold, pp. 55–63. Buffalo, N.Y.: SUNY, Canadian-American Center.

Ohline, B.A. 1987. Clean Air Act–Transboundary acid rain pollution abatement–Administrative discretion citizen suit. Nat. Resour. J. 27:707–22.

Research Consultation Group. 1979. The LRTAP Problem in North America: A Preliminary Overview. Group report. U.S. Dep. of State and Can. Dep. of Ext. Affairs.

Research Consultation Group. 1980. Second report on the Long Range Transport of Air Pollutants. U.S. Dep. State and Can. Dep. of Ext. Affairs.

United States–Canada Memorandum of Intent on Transboundary Air Pollution. 1980. Various work group reports.

Wetstone, G.S. 1984. National recourses for international pollution: Towards a United States, Canada solution. J. Air Pollut. Control Assoc. 34:111–18.

CHAPTER 2

Sources of Acid Rain

Introduction

The term "acid rain," which appears in conventional news media discussions, refers to a complex phenomenon in which sulfur dioxide (SO_2) and the oxides of nitrogen (denoted collectively by NO_x) are oxidized in the atmosphere and the resultant compounds fall to earth in rain, snow or hail, or as dry particles and aerosols (Fig. 2.1). Since acidic compounds are formed after primary pollutants (SO_2 and NO_x) are emitted, acid rain is considered to be a secondary pollutant. The primary pollutants are referred to as "precursor pollutants" of acid rain. In strict scientific terminology, "acid rain" refers to rainfall that has an acidity level beyond that which is expected in non-polluted rainfall. The term "acid precipitation" is used to include acidic snowfall and hail as well as acid rain. The term "acid deposition" is used to include the dry deposition of acidic compounds (in gaseous and particulate form) as well as the wet deposition of acidic compounds in acid precipitation. Recently, scientific discussion has expanded to consider "atmospheric deposition," which includes acidic compounds as well as other airborne pollutants. Atmospheric deposition recognizes that air pollution involves the complex interaction of a variety of compounds in a chemical "soup" in the atmosphere. While in this book the emphasis is on acid rain in the popular sense (i.e., including wet and dry deposition), other compounds in atmospheric deposition will be discussed where relevant to the acid rain debate. Chief among these other pollutants is ozone (O_3). Ozone is a secondary pollutant

2: Sources of Acid Rain

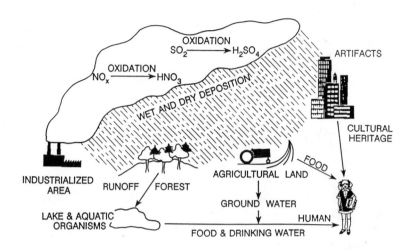

Fig. 2.1. Schematic representation of the acid deposition phenomenon. (United States–Canada Memorandum of Intent on Transboundary Air Pollution.)

formed by the photochemical oxidation of nitrogen dioxide (NO_2) in the presence of hydrocarbons (HC). NO_2 is a precursor pollutant for both acid rain and ozone.

In the following chapters it is important to keep in mind that acid rain and ozone have a common precursor pollutant. Thus, reducing ozone precursors *also* reduces acid rain precursors. The air pollution debate frequently treats these pollution problems as separable and distinct phenomena.

Since the strict definition of acid rain refers to rainfall that is more acidic than non-polluted rainfall, it is important to determine a threshold between acid rain and non-acid rain. The acidity of a solution is measured by its pH value, which measures the hydrogen ion content of the solution. The pH scale ranges from 1 (very acidic) to 14 (very alkaline). A pH value of 7 indicates a neutral solution, i.e., a solution that is neither acidic nor alkaline. The pH scale is logarithmic; hence, each unit decrease in pH represents a ten-fold increase in acidity. For example, a solution with a pH of 6 is 10 times as acidic as a solution with a pH of 7.

It is generally acknowledged that "clean rain" is slightly acidic (even in the absence of pollution), due to naturally occurring carbon dioxide (CO_2). The Research Consultation Group suggested in their first report (RCG 1979) that non-polluted rain in eastern North America would be expected to have a pH of 5.6. Rainfall with a pH <5.6 would be termed "acid rain."

While pH 5.6 is the most common pH reference point for identifying acid rain, it has not received universal acceptance (Sequeira 1982; Brocksen and Lefohn 1984). In fact pH 5.6 is the pH of laboratory distilled water in equilibrium with CO_2. Opponents argue that real non-polluted rain still has a variety of natural airborne constituents (not found in laboratory distilled water) that may be acidifying or neutralizing, causing the pH of non-polluted rainfall to diverge from the pH 5.6 value (Sequeira 1982). To deny this is to deny that rainfall could ever have had a pH in excess of 5.6, which seems to be easily refutable (Sequeira 1982). Pierson and Chang (1986) conclude that the "pH of non-polluted rainwater is likely to vary from one locality to another, with an average probably about 5. . . . There is no such thing as a single pH value for non-polluted rain and that arguments based on the proposition that there is one—5.7 or any other number—have no merit." It is important to keep this debate in mind during the discussions of impacts of acid rain in the following chapters.

SO_2, NO_x, and Acid Rain

The RCG (1979) argued that roughly two-thirds of the additional acidity in precipitation (beyond naturally occurring CO_2) is attributable to SO_2, while the remaining one-third is due to NO_x. The emissions of SO_2 and NO_x put into the atmosphere may be from natural sources or from sources within the human economic system (referred to as anthropogenic sources). On a global basis natural sources contribute roughly 60% of SO_2 emissions, while anthropogenic sources account for the remaining 40% (Pierson and Chang 1986). However, in eastern North America, anthropogenic sources from within that region account for 90% of SO_2 emissions, while natural sources in the region account for 4%; the remainder is from sources outside the region (Galloway and Whelpdale 1980). For the NO_x emissions, Pierson and Chang (1986) suggest that, while total emissions are extremely uncertain, globally natural and anthropogenic emissions are likely to be comparable in magnitude, while in eastern North America anthropogenic sources of NO_x dominate.

Natural sources may be from biogenic sources including vegetative and microbial emissions from terrestrial and oceanic regions, or nonbiogenic sources such as natural combustion, lightning, airborne soil and water aerosols, or geothermal activity (NAPAP 1985).

American and Canadian Contributions to SO_2 and NO_x Emissions

Table 2.1 shows the emissions of SO_2 and NO_x from American and Canadian anthropogenic sources in 1980, the projected emissions for 1990, and the projected emissions for the year 2000. The United States is the larger source of anthropogenic emissions of both SO_2 and NO_x in absolute terms with 24.1 million tons of SO_2 and 19.3 million tons of NO_x being generated by American economic activity in 1980. Emissions of SO_2 from American sources are forecast to decline by slightly less than 10% by 1990, while NO_x emissions increase marginally. However, from 1990 to 2000, American emissions of SO_2 are projected to increase by 16.7% and American emissions of NO_x are projected to increase by 23.5%. Canadian emissions are considerably smaller with 4.77 million tons of SO_2 and 1.83 million tons of NO_x being generated in 1980 by Canadian economic activity. Projected emissions for 1990 and 2000 in Canada show only a marginal decrease for SO_2. A marginal increase for NO_x occurs by 1990 but between 1990 and 2000, Canadian emissions of NO_x increase by 24%.

It should not be surprising that the U.S. emissions are larger than the Canadian. The U.S. economy is considerably larger than the Canadian and hence the emissions from economic activity should be

Table 2.1. Anthropogenic emissions of SO_2 and NO_X

Origin	SO_2			NO_X		
	1980	1990[1]	2000[1]	1980	1990[1]	2000[1]
	(millions of tons)					
United States	24.1	22.8	26.6	19.3	19.5	24.1
Canada	4.77	4.70	4.50	1.83	1.93	2.40

Source: Memorandum of Intent (MOI), *Emissions, Costs, and Engineering Assessment,* June 1982.

[1] Projected.

correspondingly larger. The American economy is roughly 10 times as large as the Canadian, so one might anticipate emissions in the United States to be roughly 10 times as large as those in Canada. A quick glance at the NO_x emission figures in Table 2.1 confirms this anticipation. However, this is not the case for the SO_2 emissions! Canadian emissions are roughly one-fifth that of the United States; this is twice as large as one would expect given the relative sizes of the two economies. Canada is the relatively larger contributor of SO_2 emissions—a fact that American spokespeople (industrial or political) like to point out to those Canadians arguing for acid rain controls.

Industrial Sources of SO_2 and NO_x in the United States and Canada

Whenever the relative Canadian and American performances diverge from the 1:10 ratio, it is an indicator that in this area of divergence the Canadian activity is not a simple mini-replica of the American. Table 2.2 summarizes the major activities contributing to SO_2 emissions in Canada and the United States. In the United States the major sector contributing to SO_2 emissions is the electric-power-generating industry, which accounts for two-thirds of the total. In Canada almost half of the SO_2 emissions are generated by the non-ferrous smelting industry (predominantly copper and nickel refining operations). In Canada, the electric-power industry contributes only 16% of the Canadian total, while the American non-ferrous smelting accounts for only 6% of American SO_2 emissions.

The major anthropogenic sources of NO_x in the United States and Canada are listed in Table 2.3. For both countries the predominant source of anthropogenic NO_x emissions is the transportation

Table 2.2. Major anthropogenic sources of SO_2 in the United States and Canada

	Percentage Contribution					
	United States			Canada		
Source of Pollution	1980	1990[1]	2000[1]	1980	1990[1]	2000[1]
Electric utilities	66	70	61	16	14	15
Non-ferrous smelters	6	2	2	45	49	51
Other industrial processes[2]	12	5	6	19	25	25
All other sources	16	23	32	20	11	8

Source: Calculations based upon data in MOI, *Emissions, Costs, and Engineering Assessment*, Tables A.2.1 and A.2.3, June 1982.
[1]Projected.
[2]Activities included in this category may differ between countries.

2: Sources of Acid Rain

Table 2.3. Major sources of NO_x in the United States and Canada

| | Percentage Contribution | | | | | |
| | United States | | | Canada | | |
Source of Pollution	1980	1990[1]	2000[1]	1980	1990[1]	2000[1]
Transportation	44	40	40	61	68	69
Electric utilities	29	37	36	14	10	11
All other	27	23	24	25	22	20

Source: Calculations based upon data in MOI, *Emissions, Costs, and Engineering Assessment,* Tables A.2.2 and A.2.4, June 1982.
[1]Projected.

sector. This sector is a relatively larger contributor to overall Canadian NO_x emissions than the American transportation sector. On the other hand the American electric utilities are relatively larger contributors than the Canadian. These relative contributions reflect the facts that (1) Canadian automobile exhaust standards have been less stringent (until recently) than those in the United States, and (2) the Canadian electric utilities rely on hydroelectric and nuclear power as well as fossil-fueled plants. The comparatively lax Canadian standards on automotive exhaust have been used against the Canadian government in the general debate over acid rain controls. The Canadian government tended to focus exclusively on SO_2 control for controlling acid rain while U.S. industrial and political spokespeople pointed to the lax treatment of NO_x from automobiles by the Canadian government. The International Nitrogen Oxides Protocol signed in late 1988 requires Canada (and the European signatories) to adopt regulations that are currently part of the U.S. Clean Air Act (*Coal Week* 1988).

The statistics for national SO_2 emissions by economic activity conceal important regional differences in the United States. While nationwide in 1980 the electric utilities contribute 66% of total anthropogenic SO_2, regionally the contribution differs considerably as shown in Table 2.4.

Table 2.4. Relative contributions of electric utilities and non-ferrous smelters to 1980 SO_2 emissions for selected states

	Electric Utilities	Non-ferrous Smelters
	(%)	
Illinois	77	0.00
Indiana	77	0.00
Kentucky	90	0.01
Tennessee	87	0.01
West Virginia	87	0.01
Arizona	10	86.00
Nevada	14	82.00
New Mexico	32	24.00

Source: Calculations based upon MOI, *Emissions, Costs, and Engineering Estimates,* Table B.2.1, June 1982.

The electric-power industry in the eastern United States accounts for more than 75% of SO_2 emissions from that region (as high as 90% in Kentucky). However, in the Southwest the non-ferrous smelting industries account for a large proportion of SO_2 emissions in that region (as high as 86% in the case of Arizona).

Geographical Sources of SO_2 and NO_x Emissions

Figure 2.2 reveals the geographic sources, by intensity of emissions, for SO_2 emissions in eastern North America. The darker the square the greater the emissions in that area. Two of the darkest rectangular regions appear on the Canadian side of the border. The first of these occurs just above Georgian Bay (northeastern arm of Lake Huron) and is the city of Sudbury, Ontario, where the International Nickel Company (INCO) and Falconbridge Mines are located. These companies process copper and nickel. The INCO plant is the largest single point source of SO_2 in North America, a fact that has been stressed by the U.S. spokespeople who suggest that Canada look to its own industries before looking southward for reduction in SO_2. INCO concedes that it is a large emitter of SO_2 but in its defense points out that it emits no NO_x. The second dark region on the Canadian side of the border represents the Noranda Mines at Rouyn, Quebec. This is a copper mine in the French-speaking province of Quebec and has received less adverse attention and publicity within Canada than its Ontario counterpart, INCO, perhaps for sensitive geopolitical reasons. Noranda's emissions are also relatively small compared to those of INCO (J.A. Donnan, personal communication).

The largest concentration of SO_2 emissions geographically is in the Upper Ohio River Valley–Western Pennsylvania, Indiana, Illinois, Ohio, Kentucky, West Virginia, and Tennessee where a number of coal-fired electric-power plants burn untreated, high-sulfur *local* coal. These plants have been exempted from the stringent conditions of the U.S. Clean Air Act and its amendments by virtue of their existence prior to the legislation coming into effect. Unlike these power plants, INCO has not been exempted from Ontario environmental controls and within a decade its emissions declined from between 6000–7000 tons of SO_2 per day in 1969 to about 3000 tons per day. While INCO concedes to still being North America's largest single *point* source of SO_2 emissions, it enjoys pointing out that the

2: Sources of Acid Rain

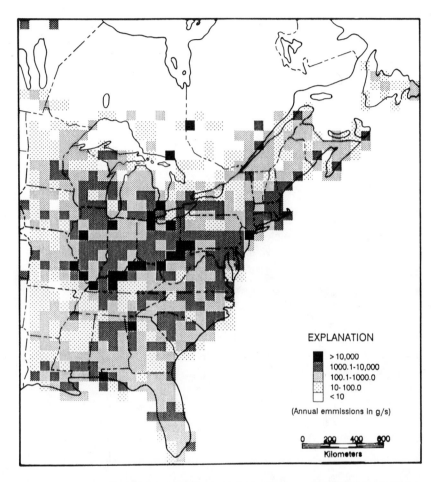

Fig. 2.2. Magnitude and distribution of sulfur dioxide (SO_2) emissions in eastern North America. (United States–Canada Research Consultation Group on the Long-Range Transport of Air Pollutants.)

largest *corporate* emitter of SO_2 in North America is the Tennessee Valley Authority (TVA).

The geographic contributions of NO_x are shown in Figure 2.3 and are relatively uniform over the industrial regions of the Ohio River Valley, the U.S. eastern seaboard, and southwestern Ontario in Canada.

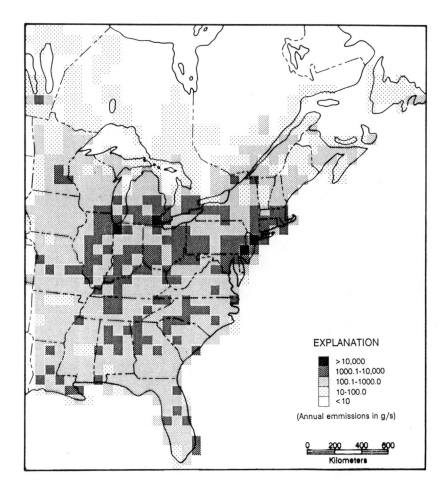

Fig. 2.3. Magnitude and distribution of emissions of nitrogen dioxide (NO_2) in eastern North America. (United States–Canada Research Consultation Group on the Long-Range Transport of Air Pollutants.)

Transport of Airborne Pollutants: Long Range or Short Range?

A major aspect of the acid rain debate among regions within the United States and between Canada and the United States as individual nations is the long-range transport of air pollutants (LRTAP) phenomenon. In terms of acid rain, what is in contention is the relative contribution of local emission sources versus emission sources outside a given region to the acidity of its rainfall.

In the LRTAP explanation of acid rain, a warm, moist air mass gathers over the Gulf of Mexico and in the summer months proceeds northward. As it moves over the industrial basin in the U.S. Midwest it picks up air pollutants, carrying them long distances where they are subsequently deposited (RCG 1979).

In part this phenomenon of long-range transport has been facilitated by the industrial response to requirements to reduce local air-pollution concentrations in the late 1960s and early 1970s. In order to meet local concentration requirements, companies built tall smokestacks to disperse the given pollutants, thereby reducing local concentrations. In 1950, more than 75% of U.S. SO_2 emissions came from stacks shorter than 100 meters. Less than 5% of U.S. emissions came from such short stacks in 1980, while close to 60% came from stacks over 200 meters in height (Altshuller 1984). In Canada in 1977–1978, roughly 63% of sulfur compounds came from stacks in excess of 200 meters high and 38% from superstacks in excess of 300 meters high (Altshuller 1984). One of the most famous (or infamous) superstacks is that of INCO (about 379 m high), which was completed in 1972 in compliance with a Ministerial Order of the Ontario Ministry of the Environment. This superstack has contributed to a dramatic improvement in air quality in and around the city of Sudbury.

The success of tall stacks in reducing local air-pollution concentrations gives rise to the potential for long-range transport by allowing the pollutants to remain in the atmosphere for prolonged periods of time and for the complex chemical reactions (required to produce the acid compounds) to take place. It is unfortunate that government policies designed to solve one environmental problem may have generated another problem.

The LRTAP explanation of acid rain is not endorsed by everyone. Poundstone (1980) believes "that a multitude of local sources will be found to be the major cause of acid precipitation, rather than long-range transport of emissions from large coal-fired power plants hundreds of miles away." A spokesperson for the Electric Power Research Institute points out that observations consistent with the LRTAP theory are also consistent with an explanation in which a meteorological condition develops and moves in a northeasterly direction converting local emissions of SO_2 into sulfate as it proceeds, and depositing the sulfate in that region rather than transporting it out of the region (Perhac 1982).

Referring to the LRTAP theory disparagingly as the "dominant theory," Anne Gorsuch (1982), former administrator of the Environmental Protection Agency (in the early Reagan administration), stated, "Despite the importance of the impact of long-range transportation of pollutants in the dominant theory of acid deposition, the finest current meteorological models do not enable us to trace emission plumes with any certainty beyond 50 k, roughly 30 miles." Rao et al. (1983) caution that

> The use of a simple trajectory model to deal with (acid precipitation) may not be appropriate. . . . It is difficult to estimate the height of the transport layer under specific meterological situations. Since the wind speed and direction vary with height, air parcel movements and pollutant concentration levels computed through the trajectory model will be highly dependent upon the transport layer chosen. In addition, under the conditions of variable flow fields, thermal advections, as well as insufficient resolution in space and time of the meterological data and variations in the underlying topography, significant errors could occur in the trajectory placements.

Research indicates local sources of sulfur emissions may be more important than long-range transportation of compounds in the northeastern United States ('News Focus' 1983). Evidence of transport from the Midwest was found only during the winter months when the levels of atmospheric sulfate are relatively low. The potential for long-range transport from tall stacks is greater in winter than summer because the atmospheric residence of sulfur is greater in the winter (Altshuller 1984).

Shaw (1982) reports that for a rural site in Nova Scotia, Canada, 50% of wet deposition of hydrogen ion and sulfate ion can be attributed to SO_2 emission sources only 25 km away in Halifax. The influence of emissions from Halifax was small for sites that were more than 100 km from Halifax. For deposition in Nova Scotia from distant sources, Shaw reports that 65% comes from the eastern United States south of the Great Lakes, 31% from the Lower Great Lakes and the St. Lawrence River Valley region, and 4% from other regions of Canada. The influence of Halifax on this site was confirmed in subsequent years (Shaw 1986).

Chan et al. (1982) determined that in data samples for August and September 1978 and June to October of 1979, the contribution of INCO emissions to wet deposition of acid and sulfur in a 50-km radius of Sudbury is relatively small and depends upon the weather system. This reflects the role of the superstack in transporting the compound away from Sudbury. For warm fronts, INCO contributes 10–20% of the deposition, while for cold fronts, the contribution is 20–40%. For summer warm fronts, the winds are from the southwest, while for cold fronts, the winds are from the northwest.

Kurtz and Scheider (1980) determined that precipitation events between August 1976 and April 1979 in the Muskoka-Haliburton region of Ontario most commonly occurred with wind trajectories from the south and southwest. Yap and Kurtz (1986) confirmed these results for 1976 through 1983 for precipitation events, wet deposition of sulfur and nitrogen compounds, and high air concentrations of sulfur and nitrogen compounds. Analysis of data collected by Scheider et al. (1980) during the shutdown of INCO between September 1978 and June 1979 showed no significant decrease in sulfate or free hydrogen-ion deposition in the Muskoka-Haliburton region. They found that the sulfate deposition level fell for sites within 12 km of Sudbury but there was no corresponding decrease in the hydrogen ion. Results of the shutdown of the INCO and Falconbridge smelters at Sudbury from June 1982 to March 1983 showed that these smelters contribute 10–20% of the total dry deposition of sulfur in central and northeastern Ontario and less than 15% of the total wet deposition in these regions (Lusis et al. 1986).

Henmi and Bresch (1985) report that acid rain episodes in the Dakotas may be attributed to emissions to the south and southeast of these states. The copper smelters in Arizona and New Mexico are likely sources. Oppenheimer (1985) concluded that 80% of Colorado's wet sulfate deposition comes from anthropogenic sources beyond Colorado's border.

Work Group 2, studying atmospheric modelling as part of the Memorandum of Intent 1980, has utilized different long-range transport (LRT) models in order to develop preliminary transfer matrices. The elements of these matrices relate emissions from a given source region to deposition in a given receptor region.

Table 2.5 presents a subset of the 60 × 60 transfer matrix generated using the Advanced Statistical Trajectory Regional Air Pollution Control (ASTRAP) Model for sulfur deposition. Most existing models predict sulfur deposition rather than acidity or nitrate deposition; however, Shannon and Lesht (1986) use the ASTRAP model to predict nitrogen deposition. Table 2.5 shows the three largest U.S. SO_2 source areas, the largest Canadian SO_2 source area, and 10 highly sensitive receptor regions (Maine is assumed to be the same as southern Nova Scotia).

Table 2.5. Total annual sulfur deposition

Sensitive Receptor Regions	Selected Major Source Areas			
	Southern Indiana	Southern Ohio	Southern Michigan	Sudbury
	($kg/ha^{-1}/yr^{-1}$)			
New Hampshire	0.63	1.30	1.60	1.00
Adirondacks	0.91	2.00	2.50	1.30
Pennsylvania	2.30	9.00	2.80	0.15
Southern Appalachia	2.20	2.20	0.17	0.01
Florida	0.08	0.06	0.01	0.00
Arkansas	0.38	0.15	0.06	0.00
Boundary Waters	0.11	0.11	0.20	0.01
Ontario	1.10	2.00	5.10	6.40
Quebec	0.61	1.10	2.20	3.50
Southern Nova Scotia	0.43	0.88	1.10	0.83

Source: Work Group 2, MOI, *Interim Report,* Table 5.1, February 1981.

The statistics for Sudbury indicate the deposition received in the receptor regions which is attributed to emissions at Sudbury (where INCO and Falconbridge Mines are located). Data indicate the deposition in Ontario is from the selected source areas. Notice that the model predicts that the bulk of Sudbury emissions falls within Canadian borders, and in terms of Ontario and Quebec, the deposition from Sudbury is less than, or about equal to, that received from the three major U.S. areas. However, this table is only a subset of the total.

Different LRT models produce different predictions concerning the amounts of deposition. Table 2.6 presents the predictions from LRT models as well as observed values. Work Group 2 (1981) suggests that the variations in predictions in Table 2.6 "are due to many differences such as: the variations in emissions inputs; the differing meteorology in years chosen to run the models; the differences in the value chosen for SO_2 to SO_4 conversion rates and wet and dry deposition."

In an early estimation of sulfur emissions, deposition, and transboundary net flows between Canada and the United States, Galloway and Whelpdale (1980) estimated that 50% of anthropogenic sulfur

2: Sources of Acid Rain

Table 2.6. LRT model predictions and observed deposition of sulfur

Sensitive Areas	Model Prediction				Observed Values
	Canada		United States		
	MOE	AES	ASTRAP	RCDM	
	(kg/ha^{-1}/yr^{-1})				
Boundary Waters	2.6	1.5	<5	5	6
Algoma, Ontario	4.7	10.4	10	17	10
Muskoka, Ontario	7.1	17.6	22	20	18
Quebec-Montmorency	5.9	9.0	15	13	20
Southern Nova Scotia	6.8	5.9	5	6	12
New Hampshire	7.9	13.1	15	13	9
Adirondack-Whiteface	8.3	15.7	19	18	12
Pennsylvania-Penn State	17.2	33.5	>25	26	19
Southern Appalachians	7.4	16.7	9	18	12

Source: Work Group 2, MOI, *Interim Report,* Table 5.3, February 1981.
Note: MOE = Ontario Ministry of the Environment; AES = Atmospheric Environmental Service; RCDM = Regional Climatological Dispersion Model; ASTRAP = Advanced Statistical Trajectory Regional Air Pollution.

deposited in Canada is from sources in the United States. They further suggested that the inflow of sulfur from the United States to Canada is about 3 times as large as the amount that flows from Canada to the United States. In another study, Fay et al. (1985) estimate a larger relative flow to Canada from U.S. sources than Galloway and Whelpdale (1980). According to Fay et al. (1985), Canada receives 8.4 times more sulfur from the United States than it sends to the United States.

The Linearity Hypothesis

The LRTAP models assume that sulfate deposition is linearly related to SO_2 emissions (Fay et al. 1985). That is, the basic transfer equation can be expressed as

$$D_{ij} = f_{ij} S_j$$

where D_{ij} is sulfate deposition in receptor region i from source region j; S_j is the SO_2 emission rate in j; and f_{ij} is the fraction of 1 ton of SO_2 from region j that is ultimately deposited as sulfate in i.

The linearity assumption is important not only for apportioning the resultant deposition in modelling the LRTAP phenomenon, but it is also a crucial aspect of the policy recommendations calling for specified emissions reductions to meet specified deposition target levels in specified receptor regions. This linearity assumption (or presumption) is also subject to debate. Altshuller (1984) suggests that the lack of agreement on this issue, at the theoretical level, may be due to different interpretations of the term "linearity." The basic equation relating sulfate formation (r_s) to SO_2 may be expressed as

$$r_s = k_s \cdot SO_2$$

where k_s is a proportionality factor. If k_s is completely independent of SO_2 then the relationship between sulfate formation and SO_2 is linear. A given percentage decrease in SO_2 emissions will bring about a proportionate decrease in sulfate formation. The k_s term, however, is a variable that depends upon those oxidizing agents present that affect the efficiency of SO_2 conversion to sulfate. Altshuller (1984) argues that an indirect non-linearity exists because of the correlation between SO_2 emissions and those of NO_x and hydrocarbons (HC), which will produce oxidizing agents.

Empirical studies by Oppenheimer et al. (1985) and Epstein and Oppenheimer (1986) suggest that the relationship between smelter emissions in the western United States and precipitation sulfate is in fact linear. Hidy (1986) on the other hand argues that it is premature to conclude that the western relationship is linear.

Indicators of Acid Rain

The discussion in the last few pages has concentrated upon (predictions of) sulfur or sulfate deposition. In these discussions the sulfur compounds are being used as a surrogate for acid deposition. The use of sulfur deposition as a surrogate for acid deposition has been questioned. If pH, SO_x and NO_x compounds always occur in fixed proportions in precipitation or deposition, then the use of any one of these as a surrogate for acid rain is legitimate (Forster 1981). However, to the extent that they can vary independently, no one of them is a sufficient indicator of acid deposition in the general sense of the term.

Sequeira (1982) concluded that neither sulfur nor nitrogen compounds are reliable indicators of, nor surrogates for, acid deposition, since various neutralizing actions such as airborne arid-soil compounds will distort the relationships between acidic compounds. Wet-fall sulfate may be correlated with cations other than hydrogen (Brocksen and Lefohn 1984).

Bess (1980) reports that for Illinois and Tennessee there has been an increase in precipitation acidity without a corresponding increase in precipitation sulfates and nitrates. The increased acidity may be due to a reduction in neutralizing cations in the precipitation. This view is supported by Sequeira's (1982) suggestion that recent decreases in pH in the eastern United States may be related to a decrease in the alkali content in the precipitation relative to the acidic components. He argues that the current level of alkali dust in American precipitation is considerably lower than it was in the mid 1950s or early 1960s. According to Bess, researchers linked the decrease in cations in Tennessee to the increased use of electrostatic precipitators, which reduce the amount of acid-neutralizing particulate matter (fly ash) that is emitted.

Time Trends in Rainfall Acidity

It is frequently suggested that the acidity of rainfall has increased over time. Support for this hypothesis is taken from representative isopleths generated by Cogbill and Likens (1974) that show increases in acidity in certain areas between the two time periods (Poundstone 1980, 1982). The relevant isopleth diagrams are shown in Figure 2.4A,B. Poundstone, a vice president of the Consolidated Coal Company, argues that by selecting different years for comparison it is possible to generate opposite trends. In responding to Poundstone (1982), Cogbill (1982) asserted that he had "never published any statement that says there is increased acidity . . ." over the stations referred to by Poundstone.

References Cited

Altshuller, A.P. 1984. The Acidic Deposition Phenomenon and Its Effects: Critical Assessment Review Papers, Vol. 1. Atmos. Sci. U.S. Environmental Protection Agency, EPA-600/8-83-016AF.
Bess, F.D. 1980. Acid rain. Ecolibrium, 1–4.
Brocksen, R.W., and A.S. Lefohn. 1984. Acid rain effects research—a status report. J. Air Pollut. Control. Assoc. 34: 1005–13.

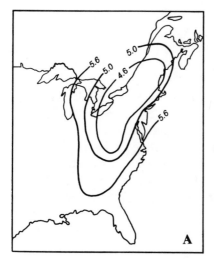 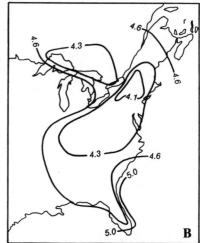

Fig. 2.4. Annual average pH isopleths: (A) 1955–1956; (B) 1975–1976. (United States–Canada Research Consultation Group on the Long-Range Transport of Air Pollutants.)

Chan, W.H., C.U. Ro, M.A. Lusis, and R.J. Vet. 1982. Impact of the INCO nickel smelter emissions on precipitation quality in the Sudbury area. Atmos. Environ. 16:801–14.

Coal Week. 1988. November 7.

Cogbill, C.V., and G.E. Likens. 1974. Acid precipitation in northeastern U.S. Water Resour. Res. 10:1133–37.

Cogbill, C. 1982. Discussion. In Acid Rain. Ed. P.S. Gold. Buffalo, N.Y.: SUNY, Canadian-American Center.

Epstein, C.B., and M. Oppenheimer. 1986. Empirical relation between sulfur dioxide emissions and acid deposition derived from monthly data. Nature 323: 245–47.

Fay, J.A., D. Golomb, and S. Kumar. 1985. Source apportionment of wet sulfate deposition in eastern North America. Atmos. Environ., 1773–82.

Forster, B.A. 1981. Separability, functional structure and aggregation for a class of models in environmental economics. J. Environ. Econ. Manage. 8:118–33.

Galloway, J.N., and D.M. Whelpdale. 1980. An atmospheric sulfur budget for eastern North America. Atmos. Environ. 14:409–17.

Gorsuch, A. 1982. The dominant theory. J. Air Pollut. Control Assoc. 32:1195–97.

Henmi, T., and J. Bresch. 1985. Meteorological case studies of regional high sulfur episodes in the western U.S. Atmos. Environ. 19:1783–96.

Hidy, G.M. 1986. Acid deposition in the western United States. Science 233:10.

Kurtz, J., and W.A. Scheider. 1980. Acidic Precipitation in South-Central Ontario: Analysis of Source Regions Using Air Parcel Trajectories, Ontario Ministry of the Environment.

Lusis, M.A., A.J.S. Tang, W.H. Chan, D. Yap, J. Kurtz, and P.K. Misra. 1986. Sudbury smelter impact on atmospheric deposition of acidic substances in Ontario. Water, Air and Soil Pollut. 30:897–908.

National Acid Precipitation Assessment Program (NAPAP). 1985. Annual Report, 1985, to the President and Congress.

'New Focus'. 1983.J. Air Pollut. Control Assoc. 33:48.

Oppenheimer, M. 1985. An analysis of sulfur budget and interstate sulfur transport for Colorado. Atmos. Environ., 1439–43.

Oppenheimer, M., C.B. Epstein, and R.E. Yunke. 1985. Acid deposition, smelter emissions, and the linearity issue in the western U.S. Science 229:859–62.

Perhac, R.M. 1982. Research program of the Electric Power Research Institute. In Acid Rain. Ed. P.S. Gold, pp. 16–23. Buffalo, N.Y.: SUNY, Canadian-American Center.

Pierson, W.R., and T.Y. Chang. 1986. Acid rain in western Europe and northeastern U.S.—a technical appraisal. CRC Critical Rev. Environ. Control 16:167–92.

Poundstone, W.N. 1980. Acid Rain: Facts vs. Allegations. Acid Rain Conference of the Energy Bureau, Inc., Arlington, Virginia, Dec. 12, 1980.

———. 1982. Acid Rain: What we know and what we don't know. In Acid Rain. Ed. P.S. Gold, pp. 70–75. Buffalo, N.Y.: SUNY, Canadian-American Center.

Rao, S.T., J. Pleim, and U. Czapski. 1983. A comparative study of two trajectory models of long-range transport. J. Air Pollut. Control. Assoc. 33:32–41.

Research Consultation Group. 1979. The LRTAP problem in North America: A preliminary overview. U.S. Dep. of State and Can. Dep. of Ext. Affairs.

Scheider, W.A., D.S. Jefferies, and P.J. Dillon. 1980. Bulk Deposition in the Sudbury and Muskoka-Haliburton Areas of Ontario During the Shutdown of INCO Ltd., in Sudbury, Ontario Ministry of the Environment.

Sequeira, R. 1982. Acid rain: An assessment based on acid-base considerations. J. Air Pollut. Control Assoc. 32:241–45.

Shannon, J.D., and B.M. Lesht. 1986. Estimation of source-receptor matrices for deposition of NO_x-N. Water, Air and Soil Pollut. 30:815–24.

Shaw, R.W. 1982. Deposition of atmospheric acid from local and distant sources at a rural site in Nova Scotia. Atmos. Environ. 16:337–48.

———. 1986. Local patterns of sulfur deposition around Halifax/Dartmouth, Nova Scotia, 1979–83. Water, Air and Soil Pollut. 30:867–72.

United States-Canada, Memorandum of Intent on Transboundary Air Pollution (MOI). 1982. Emissions, Costs and Engineering Assessment, Work Group 3B.

Work Group 2. 1981. Interim Report, Memorandum of Intent.

Yap, D., and J. Kurtz. 1986. Meteorological analysis of acidic precipitation in Ontario. Water, Air and Soil Pollut., 30:873–78.

CHAPTER 3

Agricultural Impacts

Physical Impacts

Introduction

The agricultural sector should be one of the easiest sectors for an economist to evaluate the adverse impacts of pollutants, since agricultural crops are exchanged in organized markets. Hence, data on quantities and prices are readily available, and the relevant techniques of economic analysis are well established. What is required is knowledge concerning the extent of the adverse impacts upon the supply of, or production costs of, given crops. In the case of ozone damage, much is known about the physical dose-response relationships between ozone and crop yields for many crops. In the case of acid deposition, however, establishing agreed-upon dose-response functions that demonstrate damage at current levels of deposition has proven elusive at best.

The first report on the long-range transport of air pollutants (LRTAP) of the United States–Canada Research Consultation Group (RCG 1979) claimed that "there is every indication that acid rainfall is deleterious to crops," and furthermore, there is "the potential for widespread economic damage to a number of field crops." However, there have been no reports of acid deposition–induced crop damage from actual farm operations in either the United States or Canada. Furthermore, the fears of the RCG have not been substantiated by more than a decade of research since their first report was issued.

In general, crop production, Q, is determined by the product of land area planted, A, and the yield per unit area planted, Y:

$$Q = A \times Y$$

The area to be planted for a given crop is a decision variable for each farming unit that depends upon past prices and yields of the crop, and the past prices and yields of substitute crops (Meilke and Kramar 1976).

The yield Y is influenced by (1) the above-ground environment, (2) the soil characteristics and their interactions, and (3) the genetic makeup of the plant (Miller 1983). Air pollution may reduce crop yield by adversely affecting either the aboveground environment or the soil quality once fallout of the pollution has occurred, altering soil characteristics. Impacts on crops due to a deterioration in the aboveground environment are termed "direct effects," whereas those impacts caused by a deterioration in soil quality are termed "indirect effects." The extent of damage caused to a crop may be affected by the genetic makeup of the plant.

Direct Effects of Acid Deposition and Ozone on Agricultural Crops

ACID DEPOSITION EFFECTS. Irving (1983) provides an excellent survey and critique of the research on direct effects of acid deposition on agricultural crops conducted prior to 1983. Her overall assessment of the research efforts is that "available experimental results appear to indicate that the effects of acidic precipitation on crops are minimal and that when a response occurs it may be positive or negative." The majority of crops covered in her review showed no effect of acidity on growth or yield. Huckabee (1983), in another review of experimental results, concludes that "at this time we have little indication that acidic deposition is adversely affecting crops in the United States."

The positive response to acid rain by some species has led to the hypothesis that plant response to acid deposition is determined by the interaction of an adverse impact due to low solution pH and a stimulatory response due to the presence of sulfur and nitrogen compounds (which are plant nutrients) in the deposition. According to Irving (1983) direct foliar application of these plant nutrients at critical reproductive stages may be more significant than a one-time soil fertilizer application.

In order to produce injury symptoms for crops in controlled-environment studies, the pH of the acid rain simulant must be reduced to 3.5 or below, which is almost 10 times as acidic as ambient rainfall occurring in the major crop areas (Linzon et al. 1981; Singh and Coote 1985). In field studies, the pH of the rainfall must be even lower to obtain injury symptoms (Hofstra 1986). The total deposition used in the controlled-environment studies reviewed by Irving (1983) was greater than that used in field studies reviewed. This might account for the different pH thresholds (Irving 1983). Adams and Crocker (1982) suggest that pollution impacts are greater in greenhouse experiments, since all non-pollution factors are set at biologically optimum levels, which are superior to those found under field conditions.

For the controlled-environment studies reviewed by Irving (1983), only 6 of 34 crop varieties showed any negative effects from acid rain exposure; 8 of 34 showed positive responses; 17 of 34 showed no response at all; and 3 showed both positive and negative responses. The crops in each of these categories are listed in Table 3.1.

Table 3.1. Crop responses to acid rain in controlled-environment studies

Negative Response	Positive Response	No Response	No Response	Positive and Negative Responses
Pinto bean	Alfalfa	'Wells' soybean	Rye grass	Potato
Mustard green	Tomato	Spinach	Swiss chard	'Amsoy 71'
Broccoli	Green pepper	'Limestone' lettuce	Oats	soybean
Radish	Strawberry	Cabbage	Wheat	Kidney bean
Beet	Corn	Cauliflower	Barley	
Carrot	Orchard grass	Onion	Tobacco	
	Timothy	Fescue	Green pea	
	'Oakland' lettuce	Bluegrass	Red clover	
			Bush bean	

Source: Irving (1983).

Table 3.2 lists the crop response for the 14 crop varieties studied under field conditions in Irving's review: only 1 showed a consistent negative response at all acidity levels; 3 showed a negative response for at least one acidity level; 6 showed positive yield response from at least one acidity level; and 7 showed no effect from exposure to varying levels of acidity. The negative impact on garden beets was due to fewer marketable roots per plot rather than lower root weight, and no percentage loss is given. Some crops appear in more than one column indicating a different response in different experiments either by the same researcher or different researchers. Table 3.2 also reveals that different cultivars (i.e., types) of a given crop may respond differently to acid rain.

3: Agricultural Impacts

Table 3.2. Crop responses to acid rain in field studies

Consistent Negative Response	Negative Response (at least one level)	Positive Response (at least one level)	No Response
Garden beet	'Southern Giant Curled' mustard green 'Pioneer 3992' field corn 'Amsoy' soybean	'Champion' radish 'Cherry Belle' radish 'Vernal' alfalfa 'Alta' fescue 'Beeson' soybean 'Williams' soybean	'Red Kidney' kidney bean 'Davis' soybean 'Wells' soybean 'Cherry Belle' radish 'Southern Giant Curled' mustard green 'Improved Thick Leaf' spinach 'Vernal' alfalfa

Source: Irving (1983).

The difficulty in assessing the impacts of acid rain on agricultural crops can be further understood by considering some specific examples that highlight the scientific debate between researchers in this area. Consider first the experimental evidence for yield response of radishes. Researchers have found yield increases when an acid solution containing both sulfuric and nitric acid in a 2:1 ratio is used, but yield decreases when only sulfuric acid is used, suggesting a fertilizer role for nitrogen (studies cited by Linzon et al. 1981). Such differences in experimental design will cause different outcomes. In order to address this problem, a coordinated experiment between Canadian and American research groups was undertaken. In this coordinated experiment a pH threshold of 3.0 produced a negative marketable-yield response (Province of Ontario 1983).

For soybeans the experimental results are ambiguous and have generated a sometimes heated discussion as to the basis of the conflicting results. Evans et al. (1981) found an 11.5% yield decrease in the 'Amsoy' soybean cultivar at pH 2.7 and a sulfate:nitrate ratio of 21.3. The pH is very low and the sulfate:nitrate ratio is very high. The 'Davis' cultivar showed no effect at a pH as low as 2.4 (Heagle et al. 1983) and Irving and Miller (1981) found the 'Wells' cultivar to be insensitive at pH 3.1. Yield increases of 24% and 32% were determined for 'Williams' and 'Beeson' soybeans, respectively, at pH 2.8 (Troiano et al. 1983). Irving (1983) concluded that differences in the various soybean experiments may be accounted for by differences in cultivars used, climatic conditions, and ozone concentrations. Irving suggests that the result for 'Amsoy' (Evans et al. 1981) may be attributable to toxic heavy metals and fluoride.

Evans et al. (1983, 1985b) determined yield losses of roughly

10% at pH 4.1 for 'Amsoy.' Since this pH level is close to ambient levels and given the importance of the soybean crop for North American agriculture, this finding is a cause for concern. Evans et al. (1984) defend their results of decreased yields versus the "no effect" results of Irving and Miller (1981), and the positive-effect results of Troiano et al. (1983), by claiming superior experimental design in their own work compared to Irving and Miller and Troiano et al.

Evans et al. (1984) argue that the open-top chambers (OTC) used in the Troiano work have an effect upon the plant's microclimate that can affect yield. For example, OTCs may affect the duration and/or the number of dew exposures, light transmission is reduced, and temperatures are elevated above ambient levels. The rainfall chemistry in the Troiano study lacked many constituents present in ambient rainfall as did the study of Irving and Miller (1981). On the other hand, the Evans et al. (1984) experiments had minimal impact upon the crop's micro-climate and their simulated acid rain solutions contained constituents, such as heavy metals, that are present in ambient rainfall.

In rebuttal, Irving (1984) contends that the differences may also be explained by the use of different cultivars by different research groups. In the studies discussed, only one cultivar, 'Amsoy,' showed yield losses. Even the research design of Evans and his colleagues found the cultivar 'Williams' to be insensitive to acid rain. Irving (1984) also takes issue with the Evans et al. rain chemistry, claiming that their heavy metal concentrations are much higher than those found in rural areas. To this latter point Evans et al. (1985a) respond that similar yield reductions were also obtained with lower metal concentrations. Irving (1985) counters that the yield reductions obtained with the lower metal concentrations were 3% compared to 11% in the experiment with the higher metal concentrations.

According to Adams et al. (1986) the National Acidic Precipitation Assessment Program (NAPAP) experiments during 1983 and 1984 found four of seven soybean cultivars to show significantly negative yield responses to increased acidity while two responded positively, although not significantly. The only cultivars named by Adams et al. (1986) are 'Hobbit' and 'Asgrow.' 'Asgrow' was the most sensitive cultivar with a yield reduction of 2.37% when acidity was increased by 50%. No information is given in Adams et al. (1986) as to whether 'Amsoy' was included in the NAPAP experiments.

Norby et al. (1985) found no physiological response of the

'Davis' cultivar to acidity in a greenhouse experiment. Since the greenhouse experiments generally show more sensitivity to variations in acidity than field studies, Norby et al. (1985) support the research that concludes "no soybean yield loss."

Porter et al. (1987) report on field experiments conducted over a 3-year period by 'Amsoy 71' and 'Williams 82.' For 'Amsoy 71' they determined a linear decrease in yield with increasing acidity in only 1 of the 3 years. When the most acid of the solutions was used (pH 3.0), the average yields for 'Amsoy 71' and 'Williams 82' declined by 3% and 4% respectively. Porter et al. (1987) conclude that variations in acidity of rainfall of ±50% from ambient would likely cause variations in yields for these cultivars of 1% or less.

Lee and Neely (1980) found a 9% yield decrease for field corn at pH 4.0, which is close to ambient pH in the corn-growing region. Corn is one of the most important crops in eastern North America and a 9% loss would be very significant. There was no yield loss for pH values of 3.5 or 3.0; however, the sulfate:nitrate ratio was higher for these pH levels and this may confound the results. The experiment was repeated the following year with similar, though not statistically significant, results (Irving 1983). The NAPAP experiments apparently did not find that corn was sensitive to acidity according to Adams et al. (1986). Wertheim and Craker (1988) report that in corn, pollen germination is reduced on corn silks exposed to acid rains with pH 4.6 or less compared to those exposed to pH 5.6. This decrease in pollen germination may lead to reduced corn yields, which may be particularly important when there are additional stresses such as drought or high temperature. Wertheim and Craker (1988) and NAPAP (1987a) cite studies that show yield decreases in corn when grown with simultaneous water and acid rain stresses. Dancer and Jansen (1987) find that soil acidity due to mine spoil contributes to drought stress and reduced yields for corn and soybean.

Based upon field studies of 13 cultivars across 8 different crops, NAPAP (1987a) concluded that the overall impact of acid deposition on U.S. crop production is negligible. The criticisms of the NAPAP position by other crop researchers, however, is outlined by Krupa and Lefohn (1988). The criticisms concentrated upon the number of crop species for which there was no research and the lack of adequate research into multiple combinations of temporal, physical, and chemical variables. Dry deposition effects have not been determined. The "negligible" conclusion is based upon only 2 of 8 crops that had more than one cultivar tested. The cultivar-to-cultivar variability may be as great as the variability between species. One researcher noted that the NAPAP report did not include a quantitative assessment of crop impacts. The lack of credible dose-response relationships led this researcher to conclude that the report was inadequate.

OZONE EFFECTS. The consensus among crop pathologists is that ozone is the major air pollutant causing crop damage in eastern North America (Ormrod et al. 1980; Ormrod 1981; Heck et al. 1982; Krupa et al. 1982; Pearson 1983), either alone or in combination with sulfur dioxide (SO_2) and oxides of nitrogen (NO_x).

A major coordinated research effort in the United States was mounted to determine more accurately the crop yield losses due to ozone exposure. This program, the National Crop Loss Assessment Network (NCLAN), has produced extensive information on crop losses due to ozone. NCLAN intends to provide ongoing yearly assessments to refine estimates. Research has demonstrated that visible injury (a commonly used measure of crop damage) is not always correlated with yield responses; yield losses may occur even in the absence of visible foliar injury (Heck et al. 1982).

The NCLAN research has determined yield losses for individual cultivars and has determined "common" or average responses across cultivars where it is appropriate to pool results. Table 3.3 presents the common crop responses to varying levels of ozone exposure relative to a presumed background level of 0.025 parts per million (ppm).

In addition to the crops listed in Table 3.3, the following crops have experienced some ozone injury to leaves in most years: tobacco,

Table 3.3. Percentage crop yield losses from different ozone (O_3) exposure level

Crop	O_3 Exposure (relative to 0.025 ppm)			
	0.04 ppm	0.05 ppm	0.06 ppm	0.10 ppm
Soybean	8.0	13.4	18.8	39.0
Corn	0.6	1.6	3.1	17.9
Wheat	0.8	2.3	3.6	17.0
Cotton (irrigated)	6.6	11.0	15.2	30.8
Cotton (droughted)	1.3	2.8	4.4	13.0
Kidney bean	1.8	3.1	4.8	13.2
Peanut	7.0	12.6	19.7	53.1
Lettuce	13.7	22.2	30.3	56.7
Turnip	6.9	14.1	23.9	69.7
Spinach	4.9	9.2	14.4	39.7

Source: Modified from Heck et al. (1983).

beans, onions, potatoes, grapes, tomatoes, cucumbers, squash, pumpkins, endive, and peas (Hofstra and Ormrod 1979). Visible damage due to ozone is known by a variety of names specific to the crop affected: weatherfleck for tobacco, bronzing for beans, speckle leaf for potato, and brown leaf for grapes.

Actual crop losses will vary from year to year as meteorological conditions affect the formation of ozone. Unfortunately, the conditions that are ideal for plant growth are the same conditions that make the plant most sensitive to ozone attack. These include adequate moisture, optimum fertility, and warm temperature (Hofstra and Ormrod 1979).

Indirect Effects of Acid Deposition on Agricultural Crops: Agricultural Soil Quality

Soil acidification is a natural ongoing process, which may be speeded up by acid deposition leading to a further deterioration in soil quality. According to McFee (1980), the potential effects of acid deposition on soils include (1) lower soil pH, (2) increased loss of plant nutrients, (3) accelerated weathering of mineral components, (4) decreased rates of organic matter decay, (5) changes in soil organism populations, (6) mobilization of aluminum ions, and (7) reduction of cation exchange capacity.

For maximum biomass production the soil pH should be between 5.0 and 6.2. If the soil pH is between 4.2 and 5.0, then plants in these soils will suffer periodic aluminum toxicity. For soil pH between 3.0 and 4.2, plants suffer continuous aluminum toxicity leading to a reduction in biomass production and a loss of susceptible plant species. In soils with pH < 3.5, only plants with root systems restricted to the top organic layer can survive (Arp 1983).

The degree to which a soil is sensitive to acidification depends upon its ability to neutralize the acidity, which in turn depends upon its exchangeable bases. Soils that could lose more than 25% of their exchangeable bases in 25 years are classified as sensitive, while those soils that would lose 10–25% in 25 years are classified as moderately sensitive. Soils that would lose less than 10% in 25 years are classified as non-sensitive.

The consensus of opinion among soil scientists is that acid deposition is unlikely to cause a deterioration in soil quality for properly managed agricultural lands. According to McFee et al. (1983), those studies that obtain depression in soil pH due to acid inputs used accelerated application rates or concentrated acid. There are two basic reasons why soil scientists believe that acid deposition is unlikely to cause a deterioration in properly managed croplands: the contributions of acidic input from acid deposition are small compared to the acidic contributions of fertilizer applications; the application of lime by farmers to maintain desired soil pH levels is a standard practice. For poorly managed, unamended soils or for soils growing forage crops that are not highly managed, acid deposition may have long-term effects (Irving 1983). It has also been noted that the long-term impacts upon micronutrient recycling in agricultural soils are unknown (Krupa and Lefohn 1988).

Acid deposition also has potential beneficial effects on agricultural soils, since the sulfur and nitrogen (N) compounds in acid deposition are plant nutrients. The nitrogen compounds in acid deposition are small compared to crop requirements of 100–300 kg N/ha/year (Evans et al. 1981); however, the atmospheric nitrogen inputs may be important for forage crops that do not normally receive nitrogen fertilization (D.R. Coote, personal communication). There have been no reports of sulfur deficiencies in the northeastern United States (Black 1981), nor in eastern Canada (Bixby and Beaton 1970). The contribution of atmospheric sulfur is close to the recommended sulfur requirements for crops (Evans et al. 1981). Mnkeni and MacKenzie (1981) and Gupta and MacLeod (1984) have suggested that Quebec and the Atlantic provinces, respectively, receive sufficient sulfur from atmospheric sources to meet their growth requirements.

Substantial yield responses to sulfur fertilization in the western Canadian province of Alberta have been found at an application rate of 10 kg S/ha/year. In eastern Canada the average sulfur deposition from atmospheric sources is roughly 11 kg S/ha/year (Forster 1984). For 10 principal U.S. crops, Terman (1978) calculated an average sulfur requirement of 18.5 kg S/ha/year. According to Bob Morris of the Sulfur Institute, the sulfur fertilizer market in the southeast United States is expanding in part due to declining atmospheric sulfur inputs (Crops and Soils 1986).

The net effect of acid deposition on agricultural soils depends upon the negative effects of the hydrogen ion that must be offset through lime application and the beneficial effects of free atmospheric sulfur and nitrogen that reduce fertilizer requirements.

Economic Impacts

Direct Effects of Acid Deposition

Estimates of economic losses due to the direct effects of acid deposition are summarized in Table 3.4. The first comprehensive estimates of economic impacts of acid deposition for the United States were calculated by Crocker et al. (1980). The estimates for direct agricultural impacts were made prior to the extensive crop research that would be undertaken during the 1980s. In the absence of dose-response information linking crop yield changes to acid deposition, Crocker et al. (1980) drew upon the results of oxidant damage to southern California crops, which showed a 3.01% decline in the sum of producer rents and consumer surpluses. Using this estimate for eight major field crops and fruit and ornamental crops in the eastern United States, Crocker et al. (1980) estimate a value of $1 billion ($ 1978). In reviewing this estimate, Crocker (1985) concluded that, in light of recent research developments, this estimate is "probably wrong"; however, no alternative estimate is offered.

Table 3.4. Crop losses due to direct effects of acid deposition in the United States and Canada

Study	Losses
United States	
Crocker et al. (1980)	$1.0 billion ($ 1978)
Adams et al. (1986)	$142 million ($ 1980)
Callaway et al. (1986)	$20.4–$152.4 million ($ 1984)
Canada	
Forster (1984)	$105 million ($ 1980)
Forster (1987)	$2.4 million ($ 1980)
Ludlow and Smit (1987)	$43.0 million ($ 1981, Ontario only)

Adams et al. (1986) use the results of a draft report of NAPAP findings, which suggests that the soybean crop is the only crop at risk from acid deposition. For a 50% increase in acidity from a calculated base case (representing a pristine situation), the weighted average yield loss is 1.05%. Using this estimate, the direct effect of acid deposition in 1988 was a combined loss of $142 million ($ 1980) in producer and consumer surpluses.

Forster (1987) discusses several problems with the Adams et al. (1986) analysis that need to be addressed before accepting their loss estimate. In particular, the calculation of the pristine base and the percentage increase in acidity required to produce the 1988 level of deposition is not clear. Adams et al. (1986) choose a rainfall pH of 4.8 as representing the pristine environment. They justify this by claiming that this would be produced by a 50% reduction in acidity, which is a target reduction in SO_2 emissions that appears in policy discussions.

Callaway et al. (1986) report a ($ 1984) value of $152 million for a 50% increase in acidity. This appears to be an updating of the Adams et al. (1986) estimate. Callaway et al. (1986) restrict their attention to soybeans because Irving (1983) shows soybeans to be the only major crop showing consistent, negative responses to acid deposition. The discussion above shows that such a conclusion is hard to obtain from Irving (1983).

In a first study of Canadian agricultural impacts, Forster (1984) used the Lee and Neely (1980) estimate of field-corn yield loss of 9% at pH 4.0. Assuming that the entire corn crop in eastern Canada was subject to this yield loss, Forster (1984) estimated that direct damage amounted to $105 million ($ 1980). Forster (1987) argued that if the NAPAP results used by Adams et al. (1986) are correct, then the Canadian agricultural losses would drop to $2.4 million ($ 1980), which is roughly 2% of the earlier Forster (1984) estimate. Both Forster estimates assume that Canadian crop prices remain fixed, which may not be appropriate if acid deposition reductions improve yields in both the United States and Canada.

Ludlow and Smit (1987) estimate the direct impacts on Ontario agriculture by considering the impacts on regional production. Where field-experiment data were available, they were used in favor of controlled-environment studies, and studies using sulfuric and nitric acid in combination were used in place of those using only sulfuric acid. Their major dose-response information source was Irving (1983) and they considered the responses of 17 Ontario crops to increased acidity: some show yield decreases, some show yield increases, and some show no effect. Those that have yields insensitive to acidity are considered, since production can shift from acid-sensitive to insensitive crops. Their analysis shows that if all acid rain were cleaned up, Ontario farm production would expand in value by $43 million ($ 1981). A major factor in the expanded value in their study was the gain in value of the Ontario corn crop.

Indirect Effects of Acid Deposition

The deleterious effect of acid rain on agricultural soils can be offset through liming to maintain ideal soil pH for various crops. However, this mitigation effort is not costless, and the monetary cost of the additional liming is one measure of the benefits from reducing acid deposition. Furthermore, the fertilization benefits of atmospheric inputs need to be figured into the "indirect effects" calculations.

Crocker et al. (1980) assumed that acid precipitation might increase lime requirements in the eastern United States by 5%, or 1.75 million tons. Using a cost figure of $6–$8 per ton they estimated a 1978 cost of $10.5 million to $14 million to offset the soil-acidifying potential of acid rain. Crocker et al. (1980) make no allowance for the beneficial effects of free atmospheric sulfur and nitrogen compounds.

Adams et al. (1986) suggest that the fertilizer benefit amounts to between 2 and 5 pounds of nitrogen per acre in the eastern half of the United States. It is assumed that fertilizer costs will be reduced $0.48–1.25 per acre given the passive fertilization. No reference is made to the input of sulfur from atmospheric sources. Nitrogen benefits should be small compared to sulfur benefits, since the atmospheric nitrogen is small compared to crop requirements (Evans et al. 1981). Adams et al. (1981) calculate a reduction of $233 million in fertilizer cost due to acid deposition. They assume that liming costs increase $0.07–0.21 per acre, giving a total increased cost of $23 million for liming. The combined effects (direct plus indirect) in this study produce a net benefit of $48 million to U.S. agriculture from acid deposition. All figures in this paragraph are 1980 dollars.

Callaway et al. (1986) assume an increase in liming costs of $0.17–0.52/ha and a reduction of $1.19–3.21/ha in fertilizer costs. They do not explain the basis of these ranges nor do they indicate the areas to which they are to be applied. The combined effect of these (to be inferred from their calculations) is a net benefit of $203.9 million ($ 1984) to U.S. agricultural soils. This, combined with their direct loss of $152.4 million, leads to a net benefit estimate of $51.5 million for U.S. agriculture from acid deposition. These are 1984 dollars, which are likely to be simple, updated calculations from Adams et al. (1986).

NAPAP (1987a) suggested that acid deposition supplies roughly 1% of the required nitrogen fertilizer during the growing season; i.e., roughly 1.5 kg N/ha. Using 1985 weighted average costs of nitrogen in fertilizers, the NAPAP (1987a) report calculated a benefit to soils of roughly $200 million due to atmospheric nitrogen. This estimate may be conservative, since (1) it does not include the nitrogen in ammonium and dry deposition, (2) foliar applications may be more efficiently used than one-time soil application, and (3) deposition outside the growing season may also be beneficial. The NAPAP (1987b) summary, however, states a value of $100 million for the atmospheric nitrogen! NAPAP (1987a) notes that atmospheric sulfur is also beneficial to crops and has economic benefit, but no benefits estimate is given, suggesting instead that "the dollar value is difficult to calculate because it is not known what percentage of that deposited is actually utilized by each crop." For liming requirements, NAPAP notes that to neutralize the wet deposition of hydrogen ion over croplands would require about 0.5% of total lime being applied on these lands. The estimated cost of this lime treatment is about $1 million. NAPAP concludes that a net benefit to U.S. soils is "probable."

For Canadian agricultural soils, Forster (1984) assumed that only those croplands classified as sensitive and moderately sensitive would require additional lime applications. Given regional estimates of atmospheric acid inputs to eastern Canadian soils, Forster calculated that roughly 56,000 metric tons of additional lime would be required. The cost of lime, including application, was roughly $20 per metric ton; hence the 1981 cost of neutralizing atmospheric acid on eastern Canadian croplands would have been $1.12 million. Forster (1984) also calculated benefits of $3.7 million from free atmospheric sulfur deposited on eastern Canadian croplands at a rate of 11 kg S/ha/year. Thus, according to Forster (1984), eastern Canadian soils receive a net benefit from acid deposition of $2.6 million. The net impact of acid deposition depends upon the appropriate estimate of direct damages from the two Forster (1984, 1987) studies or the Ludlow and Smit (1987) study.

Ludlow and Smit dismiss indirect effects upon Ontario soils because most agricultural soils in that province are calcareous and capable of neutralizing acids, and the input of atmospheric acid is negligible compared to fertilizer applications. This ignores variations in sensitivity of soils in the province. Roughly 10% of Ontario soil samples analyzed at the University of Guelph soil-testing lab require lime. For these the average recommendation is 5 metric tons/ha. The authors make no reference to the benefits to Ontario soils from atmospheric fertilization.

McLean (1981) claims that over half of Quebec soils have a pH <5.5. He estimates that acid deposition is removing 0.3 million metric tons of limestone from Quebec agricultural soils. The cost of replacing this is roughly $6 million—six times the Forster (1984) estimate for the entire eastern portion of Canada!

Effects of Ozone

As discussed above, the consensus of crop pathology research is that ozone is the major air pollutant causing damage to a variety of crops in North America. A number of studies of changes in ozone have considered regional economic impacts in the United States. Page et al. (1982) report economic damages to crops in the Ohio River Basin, drawing on "probable" yield reductions from the Ohio River Basin Energy Study (ORBES) (Loucks and Armentano 1982). Page et al. (1982) consider only producer losses. Adams et al. (1982) estimate the economic benefits of reducing California (1976) ozone levels for 14 annual crops. These benefits would amount to $46 million, given yield-loss estimates from Larsen and Heck (1976), which were based on a review of earlier studies.

Heck et al. (1983) use the NCLAN data to estimate economic damages to soybeans, corn, and wheat in the Corn Belt attributed to ozone exposures. For a reduction in ozone concentrations to 0.04 ppm, the economic benefit ranges from $0.73 billion to $1.19 billion, depending upon the mathematical specification of the dose-response function.

Adams et al. (1986) consider a national economic assessment of ozone crop damages in the United States. The model estimates economic (consumer plus producer) surplus ($ 1980). The dose-response information for crops was obtained from the NCLAN results. To eliminate ozone exposures above background levels would require an approximate 40% reduction in ozone from ambient levels. For a reduction of this size, Adams et al. (1982) estimate a benefit of $2.52 billion to U.S. agriculture.

In Canada, the only region that has experienced agricultural damages from ozone exposure over the growing season is southern

Ontario. An average ozone concentration of 0.04 ppm appears a reasonable assumption. Forster (1984) used the NCLAN estimates and calculated an Ontario cost of ozone exposure at 0.04 ppm of $23 million. The largest cost was to the soybean crop, followed by the corn crop. Linzon et al. (1984) estimated a range of $9–23 million as the benefits to Ontario agriculture from ozone reduction. Both Forster (1984) and Linzon et al. (1984) give 1980 dollar estimates.

References Cited

Adams, R.M., and T.D. Crocker. 1982. Dose-response information and environmental damage assessments: An economic perspective. J. Air Pollut. Control Assoc. 32: 1062–67.

Adams, R.M., T.D. Crocker, and N. Thanavibulchai. 1982. An economic assessment of air pollution damages to selected animal crops in southern California. J. Environ. Econ. and Manage. 9:42–58.

Adams, R.M., J.M. Callaway, and B.A. McCarl. 1986. Pollution, agriculture and social welfare: The case of acid deposition. Can. J. Agric. Econ. 34:3–19.

Arp, P.A. 1983. Modelling the effects of acid precipitation on soil leachates: A simple approach. Ecol. Modelling 19:105–17.

Bixby, D.W., and J.D. Beaton. 1970. Sulphur-containing fertilizers: Properties and applications. Tech. Bull. 17. Sulphur Inst., Washington, D.C.

Black, C.A. 1981. Additional facts you should know about acid rain. Crops and Soils 3(8):5–8.

Callaway, J.M., R.F. Darwin, and R.J. Nesse. 1986. Economic valuation of acidic deposition damages: Preliminary estimates from the 1985 NAPAP assessment. Water, Air, Soil Pollut. 31:1019–34.

Crocker, T.D. 1985. Acid deposition control benefits as problematic. J. Energy Law Policy, 339–56.

Crocker, T.D., J.T. Tschirhart, R.M. Adams, and B.A. Forster. 1980. Methods development for assessing acid precipitation control benefits, a report to the U.S. EPA.

Crops and Soils. 1986. Soil and water. Vol. 38 (6):26.

Dancer, W.S., and I.J. Jansen. 1987. Mine acidity and low crop productivity. J. Environ. Qual., 242–46.

Evans, L.S., G.R. Hendry, G.J. Stensland, D.W. Johnson, and A.J. Francis. 1981. Acidic precipitation: Considerations for an air quality standard. Water, Air, Soil Pollut. 16:469–509.

Evans, L.S., K.F. Lewin, M.J. Patti, and E.A. Cunningham. 1983. Productivity of field-grown soybeans exposed to simulated acidic rain. New Phytol. 93:377–88.

Evans, L.S., G.R. Hendrey, and K.H. Thompson. 1984. Comparison of statistical designs and experimental protocols used to evaluate rain acidity effects of field-grown soybeans. J. Air Pollut. Control Assoc., 34:1107–14.

———. 1985a. Letter to the editor on statistical design controversy. J. Air Pollut. Control Assoc. 35:99–101.

Evans, L.S., K.F. Lewin, K.A. Santucci, and M.J. Patti. 1985b. Effects of frequency and duration of simulated acid rainfalls on soybean yields. New Phytol. 100:199–208.

Forster, B.A. 1984. An economic assessment of the significance of long-range transported air pollutants for agriculture in eastern Canada. Can. J. Agric. Econ. 32:498–525.

———. 1987. Agricultural impacts of acid deposition: Some issues to consider. Can. J. Agric. Econ. 35:241–48.

Gupta, U.C., and J.A. MacLeod. 1984. Effects of various sources of sulphur and sulphur concentrations on yields of cereals and forages. Can. J. Soil Sci., 403–9.

Heagle, A.S., R.B. Philbeck, R.G. Feicht, and R.E. Ferrell. 1983. Response of soybeans to simulated acid rain in the field. J. Environ. Qual. 12:538–43.

Heck, W.W., O.C. Taylor, R.M. Adams, G. Bingham, J. Miller, E. Preston, and L. Weinstein. 1982. Assessment of crop loss from ozone. J. Air Pollut. Control Assoc. 32:353-61.

Heck, W.W., R.M. Adams, W.W. Cure, A.S. Heagle, H.E. Heggestad, R.J. Kohut, L.W. Kress, J.O. Rawlings, and O.C. Taylor. Dec. 1983. A reassessment of crop loss from ozone. Environ. Sci. Technol. 17:573–581.

Hofstra, G. 1986. Acid rain: The effects on agricultural production. Agrologist (Summer): 6–7.

Hofstra, G., and D. Ormrod. 1979. Air pollution and crop damage. In Highlights of Agricultural Research in Ontario, Vol. 2.

Huckabee, J. 1983. Effects of acidic deposition on agricultural crops. EPRI J. (July–Aug.): 51–53.

Irving, P.M. 1983. Acidic precipitation effects on crops: A review and analysis of research. J. Environ. Qual. 12:442–53.

———. 1984. Commentary. J. Air Pollut. Control Assoc. 84:1114–15.

———. 1985. Reply to Evans et al. 1985a letter to the editor on statistical design controversy. J. Air Pollut. Control Assoc. 85:159.

Irving, P.M., and J.E. Miller. 1981. Productivity of field-grown soybeans exposed to acid rain and sulphur dioxide, alone and in combination. J. Environ. Qual. 10:473–78.

Krupa, S.V., and A.S. Lefohn. 1988. Acidic precipitation: A technical amplification of NAPAP's findings. J. Air Pollut. Control Assoc. 38:766–76.

Krupa, S.V., G.C. Pratt, and P.S. Teng. May 1982. Air pollution: An important issue in plant health. Plant Dis. 66:429–34.

Larsen, R.I., and W.W. Heck. 1976. An air quality data analysis for interrelating effects, standards, and needed source reductions. 3. Vegetation injury. J. Air Pollut. Control Assoc. 26:325–33.

Lee, J.J., and G.E. Neely. 1980. CERL-OSU acid rain crop study progress report. Air Pollution Effects Branch, Corvallis Environ. Res. Lab. U.S. EPA, Corvallis, Oreg.

Linzon, S.N., R.G. Pearson, W.I. Gizyn, and M.A. Griffith. 1981. Terrestrial effects of long-range pollutants-crops and soils. In Proceedings of the APCA Conference. Acid Deposition, Knowns and Unknowns: The Canadian Perspective. Montreal, Quebec, April 7-8.

Linzon, S.N., R.G. Pearson, J.A. Donnan, and F.N. Durham. 1984. Ozone effects of crops in Ontario and related monetary values. ARB-13-84-Phyto, Ontario Ministry of the Environment.

Loucks, O.L., and T.V. Armentano. 1982. Estimating crop yield effects from ambient air pollutants in the Ohio River Valley. J. Air Pollut. Control Assoc. 32:146-150.

Ludlow, L., and B. Smit. 1987. Assessing the implications of environmental change for agricultural production: The case of acid rain in Ontario, Canada. J. Environ. Manage. 25:27-44.

McFee, W.W. 1980. Effects of atmospheric pollutants on soils. In Polluted Rain. Ed. T.Y. Toribara and M.W. Miller. New York: Plenum Press.

McFee, W.W., F. Adams, C.S. Cronan, M.K. Firestone, C.D. Foy, R.D. Harter, and D.W. Johnson. 1983. Effects on soil systems. In The Acidic Deposition Phenomenon and Its Effects. Critical assessment review papers, U.S. EPA.

McLean, R.A.N. 1981. Statement on terrestrial and aquatic effects. In Proceedings of the APCA Conference. Acid Deposition, Knowns and Unknowns: The Canadian Perspective. Montreal, Quebec, April 7-8.

Meilke, K.D., and R.E. Kramar. 1976. Acreage response in Ontario. Can. J. Agric. Econ. 24:51-66.

Mnkeni, P.N.S., and A.F. MacKenzie. 1981. Effects of sulphur and phosphorus fertilization on the yield and quality of barley in three southwestern Quebec soils. Can. J. Soil Sci. 61:351-59.

National Acidic Precipitation Assessment Program (NAPAP). 1987a. Interim Assessment Report. Vol. 4, Effects of Acidic Deposition. Chap. 6, Effects on agricultural crops. Washington, D.C.

_____. 1987b. Interim Assessment Report. Executive Summary. Washington, D.C.

Norby, R.J., D.D. Richter, and R.J. Luxmore. 1985. Physiological processes in soybean inhibited by gaseous pollutants but not acid rain. New Phytol. 100:79-85.

Ormrod, D.P. 1981. Air pollution as a disease agent. Can. J. Plant Pathol. 3:260-62.

Ormrod, D.P., J.T.A. Proctor, G. Hofstra, and M.L. Phillips. 1980. Air pollution effects on agricultural crops in Ontario: A review. Can. J. Plant Sci. 60:1023-30.

Page, W.P., G. Arbogast, R.G. Fabian, and J. Ciecka. Feb. 1982. Estimation of economic losses to the agricultural sector from airborne residuals in the Ohio River Basin region. J. Air Pollut. Control Assoc. 32:151-54.

Pearson, R.G. 1983. Summary of foliar assessment surveys for oxidant injury to field crops in southern Ontario: 1971–1982. Ontario Ministry of the Environment, Toronto.

Porter, P.M., W.L. Banwart, J.J. Hassett, and R.L. Finke. 1987. Effects of simulated acid rain on yield response of two soybean cultivars. J. Environ. Qual. 16:433–37.

Province of Ontario. 1983. A Presentation to the Michigan air pollution control commission in opposition to the consumers power company request to delay bringing its J.H. Campbell and B.C. Cobb power plants into compliance with the Michigan 1% or equivalent sulfur in fuel rule. Grand Haven, Mich.

Research Consultation Group. 1979. The LRTAP problem in North America: A preliminary overview. U.S. Dep. State and Can. Dep. Ext. Affairs.

Singh, S.S., and D.R. Coote. 1985. Acid rain: The effect on agriculture. Agrologist (Spring): 23–24.

Terman, G.L. 1978. Atmospheric sulphur—the agronomic aspects. Tech. Bull. 23. Sulfur Inst., Washington, D.C.

Troiano, J., L. Colavito, L. Heller, and D.C. McCune. 1983. Effects of acidity of simulated rain and its joint action with ambient ozone on measures of biomass and yield in soybeans. Environ. Exp. Bot. 23: 113–19.

Wertheim, F.S., and L.E. Craker. 1988. Effects of acid rain on corn silk and pollen germination. J. Environ. Qual. 17:135–38.

CHAPTER 4

Forest Impacts

Physical Impacts

Introduction

In the early 1980s it was presumed that North American forests were not only not adversely affected by acid deposition, but may have been benefiting from that deposition. By the mid-1980s this viewpoint had changed and the forest sector had become the subject of intense research efforts.

In the late 1970s and early 1980s there were no visible signs of forest damage in those regions of North America receiving substantial acid deposition. It was suggested that in fact the acid deposition may be stimulating forest growth in the short run as a result of the fertilizing effects of sulfur and, more importantly, nitrogen. Nitrogen is the limiting nutrient for many forests and hence acid deposition was viewed as supplying additional nutrients permitting an increase in forest productivity. While the acid deposition was also leaching away calcium and magnesium, these nutrients were not limiting forest growth (in other words they were in excess supply); hence, this process would not retard forest growth. However, over time the continued leaching could reduce these nutrients to levels where they would become limiting, and further inputs of acid deposition would reduce forest growth. The short-run gains from nitrogen fertilization give way to long-run losses due to the leaching of nutrient cations.

This was the position stated in February 1982 at the Peer Review of EPA-funded acid precipitation conference held in Raleigh, North Carolina. Within a year or so this position was being severely challenged by events taking place in German forests.

4: Forest Impacts

Signs of forest decline, or damage, in Germany were increasing dramatically. In 1982 a forest survey indicated that roughly 8% of Germany's forest area showed signs of damage. The 1983 survey revealed that the area showing signs of damage had increased to 34%. These declines were occurring in areas that had sulfur dioxide (SO_2) concentrations below the "no-damage" threshold levels previously thought to be safe from air-pollution damage. The possibility of an acid rain mechanism was suggested and the research effort thrust onward. The 1984 survey showed the area of decline in Germany continuing to increase, with 52% of the forested area showing signs of damage (McLaughlin 1985). One German scientist reported that in the fall of 1980, 62% of fir trees in Baden-Wurttenberg were healthy, but by the spring of 1982, only 4% were healthy (G.H.M. Krause quoted in Hileman 1984).

The trees in Germany that were most affected were the silver fir, Norway spruce, beech, and pine; however, birch, larch, maple, alder, ash, and oak were also affected (Cowling 1984; McLaughlin 1985).

Cowling (1984) documented the German situation as of the summer of 1983. The effects were seen in natural and managed forests, on trees both on fertile and infertile soils, those on basic and on acidic soils, and on all facing slopes. The major symptoms noted by Cowling (1984) include the dropping of green leaves and shoots; signs of decreased growth and abnormal allocation of photosynthesis; the killing of herbaceous vegetation below the canopies of affected trees; the development of microscopic calcium sulfate crystals in the stomates of some affected spruce trees; and abnormal shapes, sizes, orientation, and distribution of the leaves and shoots along the stem.

The fir trees showed decreased diameter growth that occurred over a 20-year period, while the spruce and beech showed decreased growth over 10- and 5-year periods, respectively. Spruce, pine, and beech trees showed "distress" crops (i.e., very heavy crops) of seeds and cones for 3 successive years. Such distress crops are common after stress impacts but generally occur only for a single year (Cowling 1984).

Coincident with the German experience, and thus adding to the alarm, dieback was noted in the northeastern United States and Canada. The relevant species was the red spruce, and the regions included the Adirondack Mountains, the Green Mountains of Vermont, and the Tremblant region of the Laurentian Mountains of Quebec (Tomlinson 1983). In 1984, Ontario maple syrup producers alleged that recent declines in their sugar maple trees may be due to acid rain. Similar declines in sugar maple resources have been noted in Quebec, Vermont, and New York (Borie 1987). Forest surveys in Quebec showed that 28% of the maple stands exhibited decline symptoms in 1984, increasing to 52% in 1985 (Lanken 1987). Des Granges (1986) suggested that this figure had increased to 85%.

Despite the excitement, concern and intensive research effort on the part of the academic community and various government research agencies during the 1980s, the forest products industry itself saw no reason to believe that acid rain posed any threat to forest resources. A possible explanation for this belief is that the losses to the forest sector caused by fire, insects, and disease are large compared to suspected damage attributable to acid deposition. For example, in Canada, in 1980–1981 the average annual loss of commercial timber from these so-called "natural" causes amounted to between 50% and 66% of net production levels (Crocker and Forster 1986). The forest products industry at the end of 1987 still believed that acid rain controls were not the answer to any problems concerning the health of North American forests (Haines 1987). On the other hand the maple sugar industry has called for air-pollution laws in the United States and Canada (Borie 1987).

Competing Hypotheses of Forest Damage in North America and Europe

The major competing explanations that have been put forward to explain the observed forest declines have been summarized by Cowling (1984) and further discussed by McLaughlin (1985). The six major hypotheses considered may be summarized as (1) attack by gaseous pollutants, (2) acid deposition–induced soil acidification, producing heavy-metal (aluminum in particular) toxicity, (3) acid deposition–induced foliar deficiency of magnesium, (4) acid deposition contributions to excess nitrogen, (5) general stress, and (6) drought.

The major gaseous pollutant of concern in both the United States and Germany is ozone because it occurs at phytotoxic levels over Europe and the United States. SO_2 may also be a concern in Europe (McLaughlin 1985). The effects of gaseous pollutants are due to direct damage to tree foliage. Damage to some forests in the United States attributable to ozone is well known, but Cowling (1985) believes the German evidence is less strong. Prinz (1985) suggests that ozone in combination with (acid) rain and fog plays a key role in the German situation but that ozone is the major damaging agent. Haines (1987), speaking as a representative of the forest products industry, also comes down on ozone as the prime suspect in observed forest damages in the United States. Similarly, a coal industry representative declared that if air pollution is responsible for forest damage, then it is oxidants not acid rain (Kerch 1987).

The fact that German forest damage was occurring in areas that had SO_2 concentrations below the "no-damage" threshold levels lead to the consideration of accumulated impacts due to low-level inputs of acidifying compounds over time. The aluminum toxicity hypothesis is generally associated with the research of Professor B. Ulrich and his colleagues at the University of Gottingen (Ulrich et al. 1980). Under this hypothesis, the atmospheric inputs of acid leads to soil acidification, which in turn leads to the mobilization of heavy metals, aluminum in particular. Aluminum is toxic to the fine-root system of the trees and leads to reduced uptake of nutrients and water. Prinz (1985) argues, however, that the low heavy-metal concentrations in the needles of trees showing signs of decline rule out the heavy-metal toxicity hypothesis.

Support for the heavy-metal toxicity explanation is offered by Linzon (1986) in the case of sugar maple decline in the Muskoka region of Ontario. The soils in this region are acidic and have elevated concentrations of aluminum. The declining trees had fine-root systems with higher aluminum concentrations than healthy trees and had suffered extensive root death. In an early study of the Sudbury, Ontario, area, Hutchinson and Whitby (1977) suggested that aluminum toxicity to trees in this area was being caused by highly acidic rainfall, which mobilized aluminum from the clay minerals in Sudbury soils.

The third explanation suggests that acid deposition leads to foliar deficiencies of magnesium, which had been noted in German forests. There are two different possible pathways whereby this impact occurs. First, the nutrients may be leached from the foliage as a result of direct deposition of acidic material to the forest canopy. The second pathway is indirect, operating through altered availability of magnesium in soils as a result of soil leaching.

In the fourth explanation, acid deposition contributes to excess nitrogen in the trees. One pathway in which damage occurs due to excess nitrogen is direct toxicity. Another suggested pathway is where the nitrogen creates physiological imbalances between shoots and roots. Yet another possibility is that the excess nitrogen predisposes the trees to secondary stresses (McLaughlin 1985).

Related to this last pathway for excess nitrogen impacts is the fifth general explanation, a general stress hypothesis, which proposes that air pollution acts as an inciting or initial stress that impairs the photosynthetic capacity of the foliage and predisposes the tree to other secondary stress symptoms.

The final explanation is a non-air pollution one in which the observed declines are attributed to drought or other natural conditions. Trees that are showing visible signs of decline or dieback often have tree-ring records that show the decline starting 20–30 years ago. This coincides with the drought conditions of the 1950s and mid-1960s in the affected regions. LeBlanc et al. (1987a, b) found that tree species in the Adirondack Mountains, believed to be susceptible to acid deposition, exhibited growth decreases after 1960, but the onset of this growth reduction was coincident with drought conditions and anomalous winter temperatures. Regression analysis indicated that growth reductions in the post-1960 period were correlated with climatic variables. However, these researchers found that a growth-climate model based upon pre-1961 data did not predict accurately the growth decreases that occurred after 1961, suggesting a structural change in the growth-climate relationship. The models underpredicted the growth decrease, given the climatic conditions. An air-pollution connection cannot be ruled out, but it is not the only mechanism either.

Conrad (1987) reports that fir trees on Mt. Mitchell in North Carolina are dead or dying. The Forest Service attributes these declines to the balsam woolly aphid, whose population has been expanding over time.

Declines of sugar maple stands in the Muskoka region of Ontario have occurred, coincident with severe epidemics of tent caterpillars in the late 1970s and regional drought conditions in the years 1976, 1977, and 1983 (Borie 1987).

The various air-pollution hypotheses concerning forest damage discussed above may be grouped into two major categories. First, there are those hypotheses that suggest the appropriate mechanism is direct damage to the tree. In this category the damage is caused by direct contact between air pollutants and the forest canopy, the foli-

4: Forest Impacts

age itself. This category includes the gaseous-pollutant impacts, the leaching of magnesium from foliar surfaces, and the presence of excess nitrogen in foliage from direct deposition.

The second category includes those hypotheses that postulate declines caused indirectly by air pollutants, producing changes in soil quality. This category includes the soil acidification and heavy-metal toxicity hypothesis, the leaching of magnesium from forest soils, and excess nitrogen in the soils.

Notice that in using this categorization, some hypotheses show up in both categories because of the alternative mechanism or pathway for the operation of the hypothesis. These two categories have quite different implications for forest health and the implementation of a pollution control program.

If the damage is direct damage to the forest canopy, then provided the damage is not too severe for too long, removing the air pollution removes the damage mechanism and the forest may be able to recover. If, however, the damage is indirect operating through modifications in forest-soil quality, then removing the air pollution does not remove the damage to the soil and hence does not remove the damage to the forest. The indirect effects may be irreversible or have a very slow recovery.

There does not appear to be any consensus among researchers as to which of the explanations is the dominant one. Most would agree that the observed declines are a result of a combination of these factors. What is being disputed is the relative importance and the relevant sequencing of impacts (Prinz 1985). Ulrich (1984) suggests that it may take 20 years of intensive research to fully understand the impacts on European forests. Even then the explanation for North American forests may be different from the explanation for European forests.

Cowling (1985) concedes that apart from specific known forest damage due to ozone, "all the presently available evidence indicating a possible role of airborne chemicals in the current 'declines' of forests in Europe and North America is *circumstantial*" (emphasis added). Manion (1985) argues that he cannot attribute a primary role for air pollution in the forest declines in the absence of a "consensus regarding the mechanism." The concept of forest decline is frequently attributed to Manion and hence his views on the subject are significant. Manion (1985) is quite critical of the West German forest survey as a research tool. He terms the field survey a "political tool not a biological evaluation," which was "poorly designed but has functioned very well as a political manipulation tool."

Manion also argues that the 10–20% foliage loss noted in most of the affected areas is well within the limits expected for normal healthy tree variation.

Economic Impacts

There have been very few studies that have attempted to estimate the economic damages to forests caused by acid deposition. This is not surprising given the state of the scientific debate concerning the mechanism of injury and the fact that much of the concern is for future impacts rather than existing losses.

Crocker and Regens (1985) and Crocker and Forster (1986) use a National Academy of Sciences estimate of a possible 5% yield reduction to estimate economic effects of acid deposition for U.S. and Canadian forests, respectively. Crocker and Regens (1985) calculated a loss to U.S. commercial timber industries of $.75 billion and a further $1.0 billion loss ($ 1978 U.S.) to wildlife habitat and forest recreation. Crocker and Forster (1986) calculated a loss to Canadian commercial timber industries of $197 million and a loss to Canadian wildlife habitat and recreation of roughly $1.3 billion ($ 1981 Canadian).

Callaway et al. (1986) calculated losses in the range of $.340 billion to $.510 billion ($ 1984 U.S.) by using a sensitivity analysis of U.S. forest productivity models assuming arbitrary growth reductions of 10%, 15%, and 20% due to acid deposition.

As Adams (1986) points out the above studies should be considered very "preliminary and primarily of qualitative value." After reviewing the natural science evidence on acid rain and forest impacts, NAPAP (1987) concluded that "at this time, no economic benefits to forests from further reduction in these pollutants can be calculated."

For the Quebec maple syrup industry, the Union des Producteurs Agricoles has estimated the total loss at $89 million ($ 1986 Canadian) (Penner 1986). Forster and Phillips (1987) point out that the losses to the maple syrup industry will be lessened to the extent that white birch and yellow birch trees may provide substitute sources of syrup, provided these species are less sensitive to the same atmospheric contaminants affecting the sugar maple. White birch has the widest distribution of any tree species in Canada (Allie and Jones 1987).

References Cited

Adams, R.M. 1986. Agriculture, forestry and related benefits of air pollution control: A review and some observations. Am. J. Agric. Econ. (May): 464–72.

Alli, I., and A.R.C. Jones. 1987. Sap yields, sugar content and soluble carbohydrates of saps and syrups of some Canadian birch and maple species. Can. J. For., 263–66.

Borie, L. 1987. Are sugar maples declining? Am. For. (November/December): 26–28, 66–70.

Callaway, J.M., R.F. Darwin, and R.J. Nesse. 1986. Economic valuation of acid deposition: Preliminary results from the 1985 NAPAP assessment. Water, Air, Soil Pollut., 31:1019–34.

Conrad, J. 1987. An acid rain trilogy. Am. For. (November/December): 21–23, 77–79.

Cowling, E.B. 1984. What is happening to Germany's forests? Environ. Forum (May): 6–11.

———. 1985. Effects of air pollution on forest: A critical review: Discussion. J. Air Pollut. Control Assoc. 35: 916–19.

Crocker, T.D., and B.A. Forster. 1986. Atmospheric deposition and forest decline. Water, Air, Soil Pollut. 31: 1007–17.

Crocker, T.D., and J.L. Regens. 1985. Acid deposition control. Environ. Sci. Technol. 19: 112–16.

Des Granges, J.L. 1986. Minutes and proceedings of evidence of the House of Commons Special Committee on acid rain, Ottawa, April 8, 1986.

Forster, B.A., and T.P. Phillips. 1987. Economic impact of acid rain on forest, aquatic, and agricultural ecosystems in Canada. Am. J. Agric. Econ., 963–69.

Haines, W. 1987. Air pollution and forest health: Three perspectives. Am. For. (November/December): 14–15, 55–56.

Hileman, B. 1984. Forest decline from air pollution. Environ. Sci. Technol. 18: 8A–10A.

Hutchinson, T.C., and L.M. Whitby. 1977. The effects of acid rainfall and heavy metal particulates on a boreal forest ecosystem near the Sudbury smelting region of Canada. Water, Air, Soil Pollut., 421–28.

Kerch, R.L. 1987. Air pollution and forest health: Three perspectives. Am. For., 15.

Lanken, D. 1987. We're killing our maples. Can. Geog., 17–27.

LeBlanc, D.C., D.J. Raynal, and E.H. White. 1987a. Acidic deposition and tree growth I: The use of stem analysis to study historical growth patterns. J. Environ. Qual., 325–33.

_____. 1987b. Acidic deposition and tree growth. II. Assessing the role of climate in recent growth declines. J. Environ. Qual., 334–40.

Linzon, S.N. 1986. Activities and results of the terrestrial effects program: Acid precipitation in Ontario study (APIOS). Water, Air, Soil Pollut. J. 31: 295–305.

Manion, P. 1985. Effects of air pollution on forests: A critical review discussion. J. Air Pollut. Control. Assoc. 35:919–22.

McLaughlin, S.B. 1985. Effects of air pollution on forests: A critical review. J. Air Pollut. Control Assoc. 35:512–34.

National Acid Precipitation Assessment Program (NAPAP). 1987. Interim Assessment Report. Vol. 4, Effects of Acidic Deposition. Chap. 7, Effects on forests. Washington, D.C.

Penner, K. 1986. Minutes and proceedings of evidence of the House of Commons Special Committee on acid rain, Ottawa, April 8, 1986.

Prinz, B. 1985. Effects of air pollution on forests: A critical review: Discussion. J. Air Pollut. Control Assoc. 35: 913–915.

Tomlinson, G.H. 1983. Air pollutants and forest decline. Environ. Sci. Technol. 17:246A–56A.

Ulrich, B. 1984. Effects of air pollution on forest ecosystems and waters: The principles demonstrated at a case study in central Europe. Atmos. Environ. 19:621–28.

Ulrich, B., R. Mayer, and P.K. Khanna. 1980. Chemical Changes due to acid precipitation in loess-derived soils in Central Europe. Soil Sci. 130: 193–99.

CHAPTER 5

Aquatic Impacts

Physical Impacts

Introduction

Knowledge concerning the adverse physical impacts of acid deposition is more certain and developed for the aquatic sector than any other receptor category, although much of that knowledge is qualitative rather than quantitative. Indeed, Dowd (1982) caustically suggested that lake acidification is the only documented problem related to environmental acidification. Brocksen and Lefohn (1984) counter that aquatic effects have been shrouded in confusion with much of what we know being based upon circumstantial evidence.

The first evidence of aquatic impacts of acid deposition in North America came from the Beamish and Harvey (1972) study of fish population losses in the La Cloche Mountain lakes near Sudbury, Ontario, in Canada. The losses were related to decreases in lake pH, which in turn were attributed to acidic precipitation generated by the local smelters at Sudbury. Evidence of acidified aquatic systems in the United States was provided by reports of a 1975 survey of water chemistry and fish population status in the Adirondack region of New York (Schofield 1976a,b).

The Aquatic Sector at Risk

In general the risk of acidification of a given aquatic body depends upon the intensity of the acidic loading input and the sensitivity of the receiving body to acidic inputs.

One indicator of the strength of the loading is naturally the pH of the precipitation. Using this indicator, a pH <4.5 is believed necessary to produce acidification. Schofield (1976c) reported that in the Adirondack region in September 5, 1974, through April 9, 1975, the storm-by-storm pH averaged 4.23, with a range of 3.94–4.83. The pH of precipitation measures the severity of wet deposition, but it is not a good indicator of the total magnitude of the loading to the receiving body. Dry deposition is ignored entirely.

In terms of deposition, sulfur inputs are believed to be more important than nitrogen compounds. There are two reasons for this belief. First, sulfur inputs exceed nitrogen inputs in most affected areas, and second, the nitrate ion tends to be retained by the terrestrial ecosystem more than the sulfate ion (Linthurst 1984). Thus the sulfate ion is mobile and is transported to the water body.

Work Group 1 of the Memorandum of Intent (MOI 1983) observed that no chemical or biological effects have been reported for regions receiving less than 20 kg S/ha/year. The Canadian members therefore believed that loadings of 20 kg S/ha/year would protect all but the most sensitive aquatic ecosystems. The American members of the work group declined to support this target loading in the final report. Brocksen and Lefohn (1984) suggested that using sulfate concentration in surface waters as an indicator of the effects of atmospheric acid inputs may be misleading, since the sulfate may be correlated with naturally occurring mineral sulfur rather than atmospheric-derived sulfur.

The sensitivity of the receiving body refers to the buffering ability or capacity to assimilate, or neutralize, incoming acidity. A water body is classified as sensitive to acidification if it has a low buffering capacity; that is, a low ability to neutralize acidity. Acidification of a waterway is defined as a loss of alkalinity. Thus one common measure of the sensitivity of a waterway is its alkalinity prior to the onset of acidification. Waterways that have alkalinity-neutralizing capacity (ANC) <200 μeq/L (μeq = microequivalents) are commonly classified as sensitive to acidification (Linthurst 1984). This limit is thought to include all aquatic systems sensitive to long-term acidification but may miss those systems sensitive to short-term acidification. The calcite saturation index (CSI), another common measure of buffering capacity, combines information on pH, alkalinity, and calcium concentrations (Jeans 1982). Waterways with a CSI >3 are sensitive, while those with a CSI <3 are insensitive. It is essential to note the

5: Aquatic Impacts

importance of both the intensity of deposition and the sensitivity of the receptor. Perhac (1982) tried to show the unimportance of acid rain in causing lake acidification by showing that three lakes receiving rainfall of the same pH had different resulting lake pH. Of course, the three had different sensitivities to acidification as a result of different buffering capabilities.

The surface water areas of the eastern United States and eastern Canada are classified by sensitivity and sulfur deposition rates in Table 5.1.

At 20 kg S/ha/year as the cutoff for threatening acid input, roughly 36,000 km² of surface water area in the United States are at risk from acid deposition. If those that are highly sensitive at the 10- to 20-kg range are included, then the figure rises to over 41,000 km². By comparison, using the 20-kg cutoff, Canada has over 44,000 km² at risk, which rises to over 75,000 km² if the highly sensitive group receiving 10–20 kg/ha is included.

Importantly, these figures are the areas at risk from acidification and *not* the areas that have already been acidified. NAPAP (1987) received considerable criticism for defining "acid" lakes as those for which the pH < 5.0 (Krupa and Lefohn 1988; Science 1987). Researchers note that impacts of acidification on fish life occur at pH between 5.0 and 6.0, as will be seen below. Thus the extent of acidification is greater than would be shown in NAPAP calculations. The areas that have been acidified nevertheless are small compared to the total surface water area (Linthurst 1984; Baker and Haines 1986).

Table 5.1. Eastern North America surface water area classified by sensitivity and sulfur deposition

	Sulfate Deposition	Sensitivity to Acidification			Total
		Low	Moderate	High	
			(km²)		
Eastern United States	0–10	3,950	1,240	10	5,200
	10–20	15,740	11,470	5,410	32,620
	20–40	27,400	19,970	13,940	61,310
	>40	7,620	2,190	210	10,020
Total		54,710	34,870	19,570	109,150
Eastern Canada	10–20	4,644	6,674	31,728	43,046
	20–40	6,608	16,067	27,308	49,983
	>40	660	554	408	1,622
Total		11,912	23,295	59,444	94,651

Source: Based upon figures in MOI (1983, Tables 8.1 and 8.3).

Impacts Upon Fish

The clearest evidence of the impact of acidification on aquatic ecosystems is the impact on fish species (Linthurst 1984). The best-known studies of lake acidification in North America are those for the La Cloche Mountain lakes conducted by Beamish and Harvey (1972). Losses of lake trout, lake herring, white sucker, and other species populations were related to decreasing pH. Subsequent studies of these lakes showed losses of walleye, burbot, and small-mouth bass populations. Spawning failure occurred in 1973 for the brown bullhead, rock bass, pumpkin seed, and northern pike (MOI 1983). Frenette et al. (1986) survey the impact on fisheries in the Quebec lakes.

For the Adirondack region, Schofield (1976a,b,c) observed that for some cases entire populations of brook trout, lake trout, white sucker, brown trout, and several cyprinid species had been eliminated. At elevations above 610 m, 100 lakes were devoid of fish because of acidification. Baker and Haines (1986) survey the impact on fisheries in various other U.S. regions.

Early concern centered on the complete elimination of fish species as the indicator of injury (Evans et al. 1981). The published natural science literature is replete with threshold concepts, where entire fish populations disappear at specific pH values or ranges of pH values. The thresholds for fish reproductive failure in the La Cloche Mountain lakes were determined by Beamish (1976) and are shown in Table 5.2.

Table 5.2 reveals that sensitivity to acidification is species specific. The literature generally does not quantify the fish that would be lost as the pH levels decline through the various thresholds. There seems to be little information about the possible losses of fish pro-

Table 5.2. Approximate pH at which fish in the La Cloche Mountain lakes ceased production

pH	Species
6.0–5.5	Small mouth bass, walleye, burbot
5.5–5.2	Lake trout, trout perch
5.2–4.7	Brown bullhead, white sucker, rock bass
4.7–4.5	Lake herring, yellow perch, lake chub

5: Aquatic Impacts

ductivity prior to the threshold values. According to Linthurst (1984), "unfortunately, loss of fish populations from acidified surface waters is not a simple process and cannot be accurately summarized as 'X' pH (or aluminum concentration) yields 'Y' response. The mechanism by which fish are lost seems to vary between aquatic systems and probably within a given system from year to year." NAPAP (1987) stated that "validated fish population response models are not yet available."

Hough et al. (1982) tried to quantify fish productivity losses prior to threshold pH values, as well as extinction losses by using the morphoedaphic index (MEI) (Ryder 1965). The MEI is used to predict fish yields and is the ratio of the total dissolved solids, TDS, to the mean lake depth, $\bar{Z}$.

$$MEI = TDS/\bar{Z}$$

Hough et al. (1982) replaced TDS with alkalinity so that fish yields could be calculated as a function of alkalinity (and mean lake depth). By definition as acidification proceeds, alkalinity diminishes and the impact upon fish yields can be predicted. Forster (1985) discussed the shortcomings of the Hough et al. (1982) approach despite its intuitive appeal. Not all researchers accept MEI as the most appropriate predictor of fish yields. Heimbuch and Young (1982) felt that TDS had little predictive power after lake surface area was considered. Hanson and Leggett (1982) found that phosphorus concentration and macrobenthos biomass/mean depth were better predictors of fish yields and biomass than the MEI. Prepas (1983) concluded that mean depth by itself was as good a predictor as MEI. Since these studies suggest that TDS is not a useful variable, the Hough et al. (1982) approach, which relies on alkalinity as a measure of TDS, is undermined. Hence, the link between acidification (loss of alkalinity) and fish yields is severed, and the Hough et al. (1982) projections become meaningless. Statistical analyses that use mean depth or surface area may be of little use in assessing the impacts of acidification.

Harvey and Lee (1982) calculated the number of species of fish that may have been lost in the La Cloche Mountain lakes due to acidification by deriving a regression equation that specified the number of fish species in a non-acid lake (pH >6.0) as a function of the logarithm of lake surface area. Given the surface area, the obtained equation was then used to predict the expected number of species in lakes with a pH <6.0. The difference between the predicted and observed number of species would be the number of species lost due to the pH decrease. Again, this approach concentrates on the loss of entire fish species and does not yield information on quantity or variety.

The various impacts of acidification on fish are reported by Rosseland (1986). The following discussion is based upon this excellent review of the biological research unless otherwise stated.

The main cause of the disappearance of fish populations in acidic waters is reproductive failure, although acute mortality of adult fish does occur. The acute mortality of adult fish is associated with episodic changes in water quality as a result of spring snowmelt or heavy rain in the autumn. These adult kills are attributed to toxic heavy metals such as aluminum or copper, which may increase in toxicity as pH declines. Aluminum is most toxic to fish at pH 5.0 (Haines 1981). Early life stages are more sensitive than adults of the same species and the pH threshold for species disappearance is higher than the lethal pH for acute mortality of adult fish (Haines 1981).

This is a reason for believing that fish disappearance is more attributable to reproductive failure; pH is the toxic agent at the egg stage, while aluminum is the toxic factor after the egg hatches. Mortality among these early life stages may result in a reduction of younger year classes or a total lack of certain classes. Valtonen and Laitinen (1988) report that acid stress resulted in one-third lower fecundity in perch. As reproduction fails, recruitment ceases and species may disappear entirely after an interval of time.

The reduction or elimination of year classes may reduce competition for food and as a result the remaining fish may respond with increased growth. This result is not guaranteed, however, as reduced growth of brook trout has been noted at sub-lethal pH values. Other effects on fish, short of mortality, include skeletal deformity and increased body burdens of heavy metals such as mercury (Haines 1981).

Non-Fish Aquatic Impacts

In addition to affecting fish resources directly, acidification of water bodies may affect aquatic biota, amphibians, aquatic birds, and other animals. The consensus of aquatic biology research is that acidification of water bodies results in a reduction in the number of species present, i.e., a reduction in "species diversity" (Linthurst 1984; Stokes 1986). Newcombe (1985) suggests the following reductions as a result of acidification to a lake pH of 6.5: insects 25%, molluscs 45%, sponges 20%, leeches 52%, zooplankton 20%, and rotifers 5%. Despite the agreed-upon reductions in species richness, there is no consistent relationship between pH changes and biomass or productivity changes for the phytoplankton community (Havens and De Costa 1985; Stokes 1986). Confer et al. (1983) found a loss of 2.4 species of zooplankton and 22.6 mg dry wt/m^2 per unit decrease in pH. Other experiments have found no change or increase in zooplankton standing crops. It appears that zooplankton biomass is sensitive only at pH <5.3 (Linthurst 1984).

Amphibians as a group are relatively insensitive to increased acidity, with many capable of surviving in solutions with pH ranging from 4.0 to 4.5 (Pierce 1985). However, the evidence suggests that many species suffer 50% mortality at much higher pH values, such as 5.0-7.0. Species such as the leopard frog and the spotted salamander are among the more sensitive species (Pierce 1985). Some species will suffer growth reductions or developmental abnormalities (Pierce 1985; Freda 1986). It is likely that adjustments will take place whereby acid-sensitive species are replaced by acid-tolerant species, producing a reduction in species diversity.

Aquatic birds are not likely to be affected directly by increased water acidity, but they may be affected by acidity-induced alterations in the food chain. Species richness of fish-eating birds is correlated with pH (Nilsson and Nilsson 1978). Birds most susceptible include the common loon, common merganser, and belted kingfisher (Longcore et al. 1985; McNichol et al. 1985). For example, 80% of the diet of the common loon is fish, with the remainder being crustaceans, molluscs, aquatic insects, and leeches (Barr 1973). As the food chain is disrupted, the birds may switch to other resources and other water bodies (Linthurst 1984). McNichol et al. (1985) report that the common goldeneye preferred acidified lakes, while the carrying capacity for loons and mergansers was reduced as a result of the loss of fish. They concluded that "lake acidification has varied and opposite impacts on different waterfowl species depending on feeding habits." Similar issues may arise for mammals. Birds and mammals may suffer from increased body burdens of toxic heavy metals as a result of increased fish body burdens.

Economic Impacts

Liming Costs

One approach to dealing with the acidification of surface waters is to actively pursue lake neutralization by applying lime or other alkaline substances. Ground limestone is the cheapest and most widely available neutralizing compound (Haines 1981). If the liming process is completely successful in fully restoring the waterway, then the liming cost estimates serve as upper-bound measures on the benefits to the aquatic sector from an acid rain controls program, since these costs could be avoided if the acid rain is eliminated.

Haines (1981) reported that the liming costs in New York ranged from $55 to $470/ha, with an average of $150. Using these figures all acidic lakes in New York could be limed at an annual cost of $5 million.

Studies at the Ontario Ministry of the Environment suggested a range of $50–500/ha. Forster (1985) assumed an average liming cost of $100/ha for a 3-year dose of buffering for only those highly sensitive areas in eastern Canada receiving in excess of 20 kg S/ha. This area amounted to roughly 2.8 million ha; hence, the cost of a 3-year buffering application was $280 million. A further 20% cost for monitoring costs raised the total cost to $336 million. On an annual basis, this amounted to $112 million.

Forster and Phillips (1987) pointed out that the Forster (1985) estimates ignored those highly sensitive areas receiving at least 10 kg S/ha and the moderately sensitive areas receiving at least 20 kg S/ha. Including these categories increases the surface area to roughly 7.6 million ha (Table 5.1), with application costs of $760 million, or $253 million, annually; monitoring costs must be added to this. Forster and Phillips (1987) also point out that for the highly sensitive areas receiving 10–20 kg S/ha, the $100/ha is likely to be too low, since these areas are likely to be less accessible and the liming technology will be more expensive; for example, using helicopter drops rather than barge spreading.

A study of Nova Scotia, Canada, salmon rivers concluded that liming was not an economic solution, since the liming cost far exceeded estimated benefits (Watt 1986). The aim of the liming project was to provide a 12,000-salmon enhancement over current catch. The 20-year total cost estimate, including capital costs for roads and silos,

5: Aquatic Impacts

operation costs, and monitoring, amounted to $95 million. Given this cost estimate, the average cost per restored salmon amounted to $400. The average value per landed salmon was less than $100, which falls very short of the cost figure. Watt argued that liming could still be desirable in order to maintain genetic diversity or unique salmon stocks, since he believes that the benefits of preserving genetic diversity cannot be adequately addressed using simple cost-benefit analysis. This subject will be discussed again below.

Kelso and Minns (1986) suggested that the liming costs for eastern Canadian surface waters would range from $75 million to $172 million per year, assuming that logistics were resolved and that all worries could be allayed. For these calculations it appears that they used a lower-bound estimate of $50/ha/year. The upper bound of $172 million/ha/year is for treatment of all lakes with a pH < 6. The details underlying these calculations are not presented by the authors.

Driscoll and Menz (1983) estimated the cost of liming 663 Adirondack lakes with <200 μeq/L ANC. For a 5-year lime addition, the estimated costs for accessible lakes ranged from $38 to $76 ($ 1982)/ha of surface area, while the estimated costs for remote areas varied from $453 to $673/ha. The within-group variation is attributed to the alkalinity target level. Remote, more inaccessible, areas are more expensive because helicopters must be used rather than boats. The largest cost component in the liming process is the transportation to the chemical storage area. The cost of the lime itself amounted to only 20–25% of the total cost. Excluded from the cost analysis were the transport costs for other required resources to the storage sites, program establishment and administration costs, costs of acquiring limnological data on the lakes, and any potential negative environmental impacts associated with the neutralization program. Excluding these costs, the total annual costs of a 5-year neutralization program for the 663 lakes were estimated to range between $2 and $4 million (1982). The Driscoll and Menz (1983) study did not attempt to determine a statistical cost function because of insufficient data on actual operating experiences of large-scale liming programs.

A subsequent study by Dutkowsky and Menz (1985) estimated a cost function for neutralizing acidic lakes using a sample of 547 Adirondack lakes. The estimation procedure assumes that the liming cost is to be minimized, subject to a specified alkalinity target level. The lakes were partitioned into 218 accessible and 329 remote groups based upon their proximity to roads.

For the accessible lakes the estimated cost function is (t statistics in parentheses):

$$
\begin{aligned}
C = &-32751.6 + 117.2 \text{ TALK} - 158.8 \text{ MALK} \\
&(-9.6) \quad\quad (10.3) \quad\quad\quad (-3.5) \\
&+ 6.2 \text{ (MALK)}^2 - 0.06 \text{ (MALK)}^3 + 16.8 \text{ NETWS} \\
&(10.8) \quad\quad\quad (-6.6) \quad\quad\quad\quad (72.9) \\
&+ 27.9 \text{ SURF} \quad\quad R^2 = .90 \\
&(11.5)
\end{aligned}
$$

and that for remote lakes the cost function is:

$$
\begin{aligned}
C = &-28623.2 + 48.2 \text{ TALK} + 110.8 \text{ MALK} \\
&(-7.4) \quad\quad (8.1) \quad\quad\quad (4.5) \\
&+ 2.6 \text{ (MALK)}^2 - 0.2 \text{ (MALK)}^3 + 0.001 \text{ (MALK)}^4 \\
&(4.0) \quad\quad\quad (-39.2) \quad\quad\quad (17.4) \\
&38.2 \text{ NETWS} + 75.7 \text{ SURF} + 667.6 \text{ DIST} \\
&(32.1) \quad\quad\quad (2.9) \quad\quad\quad (3.0) \\
&+ 160 \text{ ELEV} \quad\quad R^2 = .78 \\
&(3.0)
\end{aligned}
$$

with the independent variables defined as

MALK: lake ANC prior to neutralization
NETWS: total lake watershed area less the drainage area of other limed lakes
SURF: surface area of the lake
DIST: distance from chemical storage site to the lake
ELEV: elevation above sea level
TALK: target ANC level

In the estimated functions, all estimated parameters possessed the expected signs and were statistically significant at the 1% level. The R^2 shows very good fit especially considering the cross-sectional nature of the study. The equations show that the additional cost of treating remote lakes to the next highest target ANC level is more than twice that of accessible lakes.

While neutralization may be a technically feasible method of restoring chemical balance to aquatic systems, there may be biological problems. Problems that may arise include reacidification of the lake and the possible threat of aluminum toxicity to fish that have survived the original acidification of the water body (Haines 1981). Booth et al. (1986) found that whole-lake liming does not stop acid and aluminum pulses caused during spring melt, which could have adverse impacts upon near-shore spawning. On the other hand they found that whole-lake liming did improve whole-lake water quality so that trout restocking was successful. At the Mersey salmon hatchery in Nova Scotia, treatment with limestone reduced the mortality of the Atlantic salmon fry from 30% to 3% (MOI 1983).

In cases where lakes have lost fish species as a result of the acidification process, restoration requires that the liming program be accompanied by a restocking program. If the food chain has been disrupted as a result of the acidification, then the restocking may fail. Problems may also arise from exposing domestic or semi-domestic strains to wild conditions. Domestic strains may harm remaining original stocks, or domestic strains may fail as a result of genetic inbreeding causing a loss of wildness and adaptability (Fraser 1981; Saunders 1981). Fish stocked in several Ontario lakes did not survive after the lakes were neutralized (Haines 1981). Brocksen and Lefohn (1984) note that questions remain concerning toxic-metal accumulation in fish, longer-term chemical alterations, and the frequency of required liming.

In those cases where the lost fish stocks had evolved unique gene characteristics, it is impossible to recover these by restocking with alternatives. Where required and viable, the costs of restocking must be added to the costs of neutralization. Using this approach, the policymaker must realize that the project may be less than completely successful. In this case the cost of neutralization and restocking is less than the benefits of reducing acid deposition impacts on the aquatic sector. NAPAP (1987) concluded that "there is a shortage of credible information on the effects and effectiveness of mitigation techniques."

Sportfishing Losses

Economic losses to the sportfishing sector attributable to acidification in the Adirondacks have been considered by Menz and Mullen (1984, 1985). They used a 1976 survey of licensed New York State resident anglers to develop a travel cost model to estimate the demand for angling days and the reduction in visitation days that resulted from a loss of water acreage supporting fish due to acidification. The reduction in opportunities was reflected by excluding those lakes that exhibited a pH < 5.0 in post-1976 surveys.

Menz and Mullen (1984) calculated these losses using a per angler day value of $24.58 if interfishery substitution was considered, and $35.33 if there was no interfishery substitution. The total economic losses to the Adirondack recreational fishery then ranged from $1.7 million ($ 1982) with substitution considered to $1.96 million with no substitution. To allow for those lakes that had a pH between 5.0 and 6.0 that eventually may become acidified, they considered a doubling in the acreage losses. In this case the economic loss estimates range from $2.6 million with substitution to $3.2 million with no substitution. These estimates do not include losses stemming from reduced stream-fishing opportunities caused by acidification. In addition to these economic losses, the authors note that there will be a reduction in angler expenditures of $1–2 million annually.

Menz and Mullen (1985) report on the same survey but compute their figures using a revised per angler day value of $19.90 ($ 1976) and ignoring the substitution possibilities noted in their earlier paper. The estimated total net economic losses amount to $1.07 million with an additional $0.65 million being lost if those lakes with pH between 5.0 and 6.0 are included.

The NAPAP interim assessment report issued in late 1987 included an economic assessment of the damages to the Adirondack fishery. Two approaches were used: (1) a participation model and (2) a travel cost model. Both approaches considered two scenarios for acreage reduction: (a) a 3% reduction in fishable acreage and (b) a 10% reduction in fishable acreage. Each of these scenarios was coupled with two alternative assumptions about the impact of acidification on the catch rate. The first case assumed no impact upon the catch rate, while the second assumed that the average catch rate on the remaining lakes is reduced by an equal percentage. The 1986 dollar estimates for the travel cost approach range from $1 million to $13 million, while the participation model produces loss estimates that range from $2 million to $10 million (using a $32 per angler day value). The lower-bound estimates are quite close to the Menz and Mullen (1984, 1985) results if no adjustments for inflation are made. NAPAP (1987) suggests that the average losses to sportfishing would be in the vicinity of $5 million.

If the NAPAP (1987) results are extrapolated to the entire United

States, then the losses would range from $10 million to $100 million. Crocker and Regens (1985) presented an early damage estimate of $250 million for the eastern U.S. aquatic sector. This was the lowest of the major receptor damage categories they cited. In reviewing this estimate, Crocker (1985) suggested that a doubling or tripling of the estimate would be in order when taking into consideration the more recent information on aquatic impacts and the extent of acidification. This produces an estimate considerably larger than the NAPAP (1987) upper bound on national damages.

Forster (1985) estimated losses to Canadian recreational fishing using a range of per angler day values of $50–100, which produced an aggregate loss of $52–104 million ($ 1981) assuming a 1% loss in fishing days. Kelso and Minns (1986) estimate the recreational fishing losses in Ontario and Quebec to be $53.2 million based upon gross expenditure changes per angling day. To this they would add a further $90 million in possible future losses. This combined estimate is about $40 million higher than the upper bound of Forster's (1985) rough calculation. The larger losses may be attributable to the Kelso and Minns assumption that anglers who fished in previously nonacidic waters cease to fish when these same waters become acidified. This elimination of substitution possibilities increases the loss estimate.

Losses to Aquatic Recreation and Amenity Values

Forster (1985) argued that the analysis should focus on a broader aquatic recreational experience rather than limiting attention to sportfishing as such. In general, the impact of acidification on all recreational amenities should be considered, for the general recreationist may place value on living things in the ecosystem, in addition to the fishing experience. They may derive considerable enjoyment from merely being able to observe fish swimming or jumping with no intention of catching them. Scuba divers, for example, derive pleasure from observing fish and other marine life in their own habitat. The enjoyment of the diving experience is diminished by the disappearance of the fish. Canoeists and other boaters similarly derive pleasure from observing wildlife that depend upon aquatic resources. In Canada, frequent concern is expressed for the fate of the loon. Other waterfowl may also be at risk.

Swimming is another important aquatic-based recreation activity that may improve due to acidification because of the accompanying increase in water clarity. However, acidification may also have a detrimental effect on swimming as a result of the growth of benthic filamentous algae near shore. And, there is the possibility of obnoxious odors developing from the growth of certain phytoplankton species in poorly buffered lakes. Thus it is not immediately apparent what the net change in perceived swimming and recreation quality will be once these various traits are considered. It was not possible to determine a response in Appalachian swimming or boating activity as a result of varying pH levels (Nathans 1969).

Many of the feared adverse impacts have not occurred on a large scale during the 1980s. The actual economic damages to the aquatic sector during that decade is probably minimal (Mitchell 1980; Baker and Haines 1986). However, there is concern that continued acidification could lead to irreversible losses. In this case the economist needs to assess the value that society places on avoiding these future losses, rather than simply considering current losses. A major study was conducted on this problem by ARA Consultants (1982) for the Ontario Ministry of the Environment. This study has been discussed in some detail by Forster (1985) and Forster and Phillips (1987). The following discussion is based upon these papers.

The objectives of the ARA study were to (1) determine monetary values for changes in the quality of environmental amenities induced by acid deposition (or other forms of pollution); (2) determine the socio-economic factors that might explain variations in individuals' monetary valuations; (3) determine the level of awareness concerning acid deposition and other pollution problems; and (4) determine the substitution of activities that individuals would make as a result of the environmental-quality changes.

The study used face-to-face interviews with 920 individuals over the age of 18. In order to assess the values to non-users the study included 206 residents from an urban area distant from the region concerned. The remainder were interviewed in Ontario's major "cottage country." Two distinct locations were considered: the Muskoka-Haliburton region, which was thought to be at risk from acid deposition, and the Kawartha Lakes district, which was thought to be insensitive to acidification and hence not at risk despite high acid loadings. Since foreign tourists are an important part of Ontario tourism, 100 U.S. residents vacationing in these areas were surveyed.

The lengthy questionnaire obtained information concerning the distance travelled for recreational experiences and the proportion of

5: Aquatic Impacts

time spent on a variety of activities. Individuals were asked to identify substitute activities and sites and to assign importance levels to protection from pollution for fish; small animals; water birds, plants, and flowers; and land plants and flowers.

In order to determine the values people place on environmental-quality changes induced by acidification, the study used a "contingent valuation" approach, which is capable of dealing with future potential changes in environmental quality. In order to represent the process and the possible impacts of acidification (i.e., the contingencies) as accurately as possible, an "environmental-quality ladder" (Table 5.3) was developed in conjunction with Ontario Environment Ministry scientists. The descriptions are intended to describe the decline in species diversity that occurs as acidification proceeds. However, this is "aggregated" with other features of acidification.

The respondents were asked the direct open-ended question, "How much, if anything, would you pay in taxes and prices annually to protect the Ontario environment from declining from level 8 to level 4?" The responses were recorded and the respondents were then

Table 5.3. Environmental-quality ladder

10 "Unpolluted" environment—all fish, wildlife, and plants healthy and abundant. Restoration of some fish species such as trout and salmon to waters where they used to be naturally, but declined.
9 Environment with a wide variety of wildlife, plants, etc. In some parts of the province, the best sport fish no longer exist, such as walleye.
8 Ontario environmental-quality level—good fishermen would catch 10 fish in 2 days. Forests, wildflowers, and wildlife healthy and abundant.
7 Generally healthy and varied wildlife and vegetation. Some decline in sport fish numbers and types (bass). Average fishers would catch 7 fish in 2 days.
6 Wildlife and vegetation still relatively healthy and varied. Aquatic life not as abundant—fewer frogs, less waterplants, and loss of some sport fish such as trout. Fishers would catch 3 fish in 2 days.
5 Wildlife and vegetation declining—fewer others, ducks and loons. Very few fish types left in lake—yellow perch, chub, and suckers.
4 Only one kind of fish left—yellow perch. Frogs are rare and the small vegetation growth around the lake is rare. Water seems clear, and there are very few water birds (ducks, loons) and fish-eating wildlife (raccoons, otters).
3 No fish left in the lake and wildlife that depend on fish for food are gone. Water plants, except for moss, have disappeared and the lake water is crystal clear except for a green film in water.
2 The lake is very clear, but there is virtually no sign of fish-eating wildlife in the form of ducks or small animals and very few birds. The trees and shrubs seem thin and the leaves of some are spotted with brown.
1 Polluted environment—fish, wildlife, plants seems less healthy and less abundant. Some species no longer exist in the province.
0

Source: From ARA Consultants Survey for the Ontario Ministry of the Environment (1982).

asked, "Why did you choose this amount?" The questions were repeated for changes in levels from 8 to 7, 8 to 2, and from 8 to 6.

Table 5.4 shows the average bids by income group for each of the changes in environmental quality.

Forster and Phillips (1987) provide the following simple inverse demand function using the data in Table 5.4 relating willingness to pay (WTP) to the change in environmental quality ΔEQ and income Y:

$$WTP = -151.11 + 32.41\ \Delta EQ + 11.04Y \qquad R^2 = 0.80$$

The parameter estimates are all significant at the 1% level. This suggests that the average individual WTP increases $32.41 for each potential unit drop on the EQ ladder, while the WTP bid increases by $11.04 for each $1000 increase in income.

Forster (1985) presented the equation estimates for each of the interview groups using the individual data rather than the averages. In equation form these are:

All Respondents

$$WTP = -149.23 + 32.14\ \Delta EQ + 10.31Y \qquad R^2 = 0.16$$

Muskoka/Haliburton

$$WTP = -115.85 + 34.01\ \Delta EQ + 8.88Y \qquad R^2 = 0.20$$

Kawartha

$$WTP = -104.80 + 25.95\ \Delta EQ + 9.00Y \qquad R^2 = 0.12$$

Urban

$$WTP = -267.87 + 39.65\ \Delta EQ + 16.61Y \qquad R^2 = 0.19$$

U.S. Tourist

$$WTP = -192.17 + 30.54\ \Delta EQ + 10.15Y \qquad R^2 = 0.15$$

All parameter estimates are significant at the 1% level. The similarity of parameter estimates across the groups is striking. The re-

Table 5.4. Average willingness-to-pay to prevent a deterioration in environmental quality in Ontario

Income	Change in Ladder Position			
	8 to 7	8 to 6	8 to 4	8 to 2
<10,999	55	84	74	90
11,000–19,999	65	75	88	120
20,000–29,999	108	136	186	246
30,000–34,999	191	225	288	421
35,000+	301	369	528	668

Source: From ARA Consultants survey for the Ontario Ministry of the Environment (1982).

Note: The numbers in the table are average dollar values for the respective environmental quality given the income grouping.

sponses of the urban group are somewhat larger than the others; however, there still appears to be considerable consistency in structure across the groups. The R^2 is considerably lower than for the average, calculated equation reported in Forster and Phillips (1987). This is not surprising when large micro-data sets are used. There is a lot of variation in these data sets due to individual characteristics that do not show up when the data are averaged over income groups.

The ARA study did not calculate an aggregate WTP to protect the environment. This is probably wise considering the criticisms levied by Forster (1985). It would be necessary to determine if the interviewees were representative of the relevant populations. Were the urban residents also possible recreators in the relevant regions? Were the individuals stating "individual" or "household" WTP? The WTP bids may be lower than could have been determined through an "iterative-bidding" process. Thus the obtained WTP may under-estimate the true WTP values.

However, more damaging is the likelihood that individuals were giving values to protect *all* of Ontario from sliding down the EQ ladder, given the structure of this section of the survey. The impact of acidification will be much more regionally confined. One would expect a recreationist in the Kawarthas to bid quite differently if it is believed the Kawarthas will also be affected rather than just the Muskoka-Haliburton region.

Forster (1985), drawing on estimates from the U.S. literature, assumed a ballpark, household WTP estimate of $100 per year to preserve environmental quality in the affected areas. Given the population east of Manitoba and assuming an average household size of three, this amounted to eastern Canadians paying $560 million to preserve their sensitive aquatic regions from degradation due to acid deposition.

Neuman (1986) reports that surveys conducted between 1980 and 1985 show that 70% of Ontario residents and 50% of Canadian residents would be willing to pay higher taxes or higher prices in order to control acid rain. Over half of these groups would be willing to pay in excess of $100.

References Cited

ARA Consultants. 1982. Value Awareness and Attitudes Associated with Acid Precipitation in Ontario: The Amenity Value Survey. Report to the Ontario Ministry of the Environment, Toronto, Ontario, Canada.

Baker, J.P., and T.A. Haines. 1986. Evidence of fish population responses to acidification in the eastern United States. Water, Air, Soil Pollut. 31:605–31.

Barr, J.F. 1973. Feeding biology of the common loon (*Gavia immer*) in oligotrophic lakes in the Canadian shield. Ph.D. dissertation, University of Guelph, Ontario.

Beamish, R.J., 1976. Acidification of lakes in Canada by acid precipitation and the resulting effects on fish. Water, Air, Soil Pollut. 6:501–14.

Beamish, R.J., and H.H. Harvey. 1972. Acidification of the La Cloche Mountain lakes, Ontario, and resulting fish mortalities. J. Fish. Res. Board Can. 29:1131–43.

Booth, G.J., J.G. Hamilton, and L.A. Molot. 1986. Liming in Ontario: Short-term biological and chemical changes. Water, Air, Soil Pollut. 31:709–20.

Brocksen, R.W., and A.S. Lefohn. 1984. Acid rain effects research—a status report. J. Air Pollut. Control Assoc. 84:1005–13.

Confer, J.L., T. Kaaret, and G.E. Likens. 1983. Zooplankton diversity and biomass in recently acidified lakes. J. Fish. Aquatic Sci. 40:36–42.

Crocker, T.D. 1985. Acid deposition control benefits as problematic. J. Energy Law Policy 6:339–56.

Crocker, T.D., and J.L. Regens. 1985. Acid deposition control. Environ. Sci. Technol. 19:112–16.

Dowd, J.A. 1982. Costs with benefits: The rush to judgement. In Acid Rain: A Transjurisdictional Problem in Search of a Solution. Ed. P.S. Gold, pp. 92–103. SUNY, Buffalo: Canadian-American Center Publications.

Driscoll, C.T., and F.C. Menz. 1983. An estimate of the costs of liming to neutralize acidic Adirondack surface waters. Water Resour. Res. 19:1139–49.

Dutkowsky, D., and F.C. Menz, 1985. A cost function for neutralizing acidic Adirondack surface waters. J. Environ. Econ. Manage. 12:277–285.

Evans, L.S., G.R. Hendrey, G.J. Stensland, D.W. Johnson, and A.J. Francis.

1981. Acidic precipitation: Considerations for an air quality standard. Water, Air, Soil Pollut. 16:469–504.

Forster, B.A. 1985. Economic impact of acid deposition in the Canadian aquatic sector. In Acid Deposition. Ed. D.F. Adams and W.W. Page, pp. 409–37. New York: Plenum Press.

Forster, B.A., and T.P. Phillips. 1987. Economic impacts of acid rain on forest, aquatic and agricultural ecosystems in Canada. Am. J. Agric. Econ., 963–69.

Fraser, J.M. 1981. Comparative survival and growth of planted wild, hybrid and domestic strains of brook trout in Ontario lakes. Can. J. Fish. Aquatic Sci., 1672–84.

Freda, J. 1986. The influence of acidic pond water on amphibians: A review. Water, Air, Soil Pollut. 30:439–50.

Frenette, J.J., Y. Richard, and G. Moreau. 1986. Fish responses to acidity in Quebec lakes: A review. Water, Air, Soil Pollut. 30:461–76.

Haines, T.A. 1981. Acidic precipitation and its consequences for aquatic ecosystems: A review. Trans. Am. Fish. Soc. 110:669–707.

Hanson, J.M., and W.C. Leggett. 1982. Empirical prediction of fish biomass and yield. Can. J. Fish. Aquatic Sci. 39:257–63.

Harvey, H.H., and C. Lee. 1982. Historical changes related to surface water pH changes in Canada. In International Symposium on Acidic Precipitation and Fisheries Impacts in Northeastern North America Proceedings, pp. 45–54. Cornell University, N.Y.

Havens, K., and J. De Costa. 1985. An analysis of selective herbivore in an acid lake and its importance in controlling phytoplankton community structure. J. Plankton Res. 7:207–12.

Heimbuch, D.G., and W.D. Young, 1982. Another consideration of the morphoedaphic index. Trans. Am. Fish. Soc. 111:151–53.

Hough, Stansbury and Michalski, Ltd., and J.E. Hanna Assoc. 1982. An approach to assessing the effects of acid rain on Ontario's inland sports fisheries. Can. Dep. of Fisheries and Oceans, Rexdale, Ontario.

Jeans, D. 1982. Acid rain: A Newfoundland and Labrador perspective. In Acid Rain: A Transjurisdictional Problem in Search of a Solution. Ed. P.S. Gold, pp. 30–39. SUNY, Buffalo: Canadian-American Center.

Kelso, J.R.M., and C.K. Minns. 1986. Estimates of existing and potential impact of acidification on the freshwater fishery resources and their uses in eastern Canada. Water, Air, Soil Pollut. 31:1079–90.

Krupa, S.V., and A.S. Lefohn. 1988. Acidic precipitation: A technical amplification of NAPAP's findings. J. Air Pollut. Control Assoc. 38:766–76.

Langcore, J.R., D.G. McAuley, K.L. Stromborg, and G.L. Hensler. 1985. Waterbirds on Maine wetlands with low buffering capacity: A resource at risk. Paper presented to the International Symposium on Acidic Precipitation, Muskoka, Ontario.

Linthurst, R.A., ed. 1984. The Acidic Deposition Phenomenon and Its Effects:

Critical Assessment Review Papers. Vol. 2, Effects Sciences. EPA-600/8-83-016B, USEPA, Washington, D.C.

McNichol, D.K., B.E. Bendell, and R.K. Ross. 1985. Waterfowl and aquatic ecosystem acidification in northern Ontario. Paper presented to the International Symposium on Acidic Precipitation, Muskoka, Ontario.

Memorandum of Intent on Transboundary Air Pollution (MOI). 1983. Final report of the impact assessment work group. Ottawa and Washington, D.C.

Menz, F.C., and J.K. Mullen. 1984. Acidification impacts on fisheries: Substitution and the valuation of recreation resources. In Economic Perspectives on Acid Deposition Control. Ed. T.D. Crocker, pp. 135-55. Stoneham, Mass.: Butterworth.

_____. 1985. The effects of acidification damages on the economic value of the Adirondack fishery to New York anglers. Am. J. Agric. Econ. (February): 112-19.

Mitchell, C.L. 1980. Canada's fishing industry: A sectoral analysis. Canadian special publication of fisheries and aquatic sciences, No. 52. Dep. of Fisheries and Oceans, Ottawa.

Nathans, R.R. 1969. Impact of acid mine drainage on recreation and stream ecology. Appalachian Regional Commission, Washington, D.C.

National Acid Precipitation Assessment Program (NAPAP). 1987. Interim Assessment Report.

Neuman, K. 1986. Trends in public opinion on acid rain: A comprehensive review of existing data. Water, Air, Soil Pollut. 31:1047-60.

Newcombe, C.P. 1985. Acid deposition in aquatic ecosystems: Setting limits empirically. Environ. Manage., 277-88.

Nilsson, S.C., and I.N. Nilsson. 1978. Breeding bird community and species richness on lakes. Oikos 31:214-21.

Perhac, R.M. 1982. Research program of the Electric Power Research Institute. In Acid Rain: A Transjurisdictional Problem in Search of a Solution. Ed. P.S. Gold, pp. 16-23. SUNY, Buffalo: Canadian-American Center.

Pierce, B.A. 1985. Acid tolerance in amphibians. Bioscience 35:129-243.

Prepas, E.E. 1983. Total dissolved solids as a predictor of lake biomass and productivity. Can. J. Fish. Aquatic Sci. 40:92-95.

Rosseland, B.O. 1986. Ecological effects of acidification on tertiary consumers. Fish population responses. Water, Air, Soil Pollut. 451-60.

Ryder, R.A., 1965. A method of estimating the potential fish production of north temperate lakes. Trans. Am. Fish. Soc. 94:214-18.

Saunders, R.L., 1981. Atlantic salmon stocks and management implications in the Canadian Atlantic provinces and New England, U.S.A. Can. J. Fish. Aquatic Sci. 38:1612-25.

Schofield, C.L. 1976a. Dynamics and management of Adirondack fish populations. Final report for project F-28-R, Dep. of Environmental. Conservation, Albany, N.Y.

_____. 1976b. Acid precipitation: Effects on fish. Ambio 5:228-30.

_____. 1976c. Lake acidification in the Adirondack mountains of New York:

Causes and consequences. In Proceedings of the First International Symposium on Acid Precipitation and the Forest Ecosystem. Ed. S. Dochinger and T.A. Seliga. USDA Forest Service Tech. Report, NE-2.

Stokes, P.M. 1986. Ecological effects of acidification on primary producers in aquatic systems. Water, Air, Soil Pollut. 30:421–38.

Valtonen, T., and Laitinen, M. 1988. Acid stress in respect to calcium and magnesium concentrations in the plasma of perch during maturation and spawning. Environ. Biol. Fishes 22:147–54.

Watt, W.D. 1986. The case for liming some Nova Scotia salmon rivers. Water, Air, Soil Pollut. 31:775–90.

CHAPTER 6

Materials Damages

Physical Impacts

Introduction

Acid deposition and its precursor pollutants are known to cause damage to various materials although this receptor category was ignored in the two reports of the joint United States–Canada Research Consultation Group (RCG 1979, 1980). Information concerning the various impacts on materials is reviewed by Linthurst (1984) and the Memorandum of Intent on Transboundary Air Pollution (MOI) (1983). A more recent assessment is provided by the National Acid Precipitation Assessment Program (NAPAP 1987).

More is known about potential qualitative effects of acid deposition and its precursors on various materials categories than is known about actual quantitative impacts under field conditions. Table 6.1 summarizes the knowledge concerning impacts of air pollution on a variety of materials. Recent research has focussed on the metals, building stone and paints, and organic-coatings categories (NAPAP 1987). The primary pollutant causing damage to materials is sulfur dioxide (SO_2). There is, however, evidence that the protective patina of copper and its alloys may be adversely affected at pH < 4.0. For aluminum there is a suggestion that nitrogen oxides (NO_x) may be more important in influencing corrosion than SO_2.

It now appears that copper, aluminum, and bare wood are less sensitive to acid deposition than paint, zinc, building brick, carbonate stone, and possibly concrete (NAPAP 1987).

A number of factors combine to make the development of credible "damage" or "dose-response" functions difficult. First, it is difficult to reproduce actual conditions in chamber studies. This problem occurs generally in the materials category due to the length of the normal lifetimes of exposed materials. Second, in many experiments, "to produce measurable material damage in a reasonable period of time, the material is often exposed continuously to severe environmental conditions (e.g., extremely high pollutant concentrations and/or high humidity) completely unrepresentative of field conditions" (Linthurst 1984).

Franey and Graedel (1985) note that corrosion films that form on metals and alloys are generally chemical mixtures, hence, laboratory experiments must involve mixtures of corrosive constituents if they are to reproduce field results. They also note that experiments frequently use chlorine gas to induce a chloride formation. However, since chlorine gas is a rare atmospheric compound, this test has limited relevance. According to Franey and Graedel (1985), a realistic test would require carbon dioxide (CO_2), a sulfur-containing gas, a chlorine-containing gas such as hydrogen chloride (HCl), a nitrogen-containing gas such as nitrogen dioxide (NO_2), water vapor, simulated solar radiation, and particulate matter.

Measuring damage poses a difficulty. One measure used for both metals and stone is mass loss from dissolution. It is also possible to measure the weight loss caused by film erosion to paint and other surface coatings although mass loss is not a common cause of paint failure.

Applications to real structures pose several difficulties, since special features need to be considered. One of the most important factors determining material damage given a specific pollutant dose is the "time of wetness" of surface moisture. This time of wetness is affected not only by meteorological and environmental factors but also by physical characteristics of the structure being considered. These characteristics include the structure's orientation, its angle of exposure, and the shading effects of other buildings or other parts of the building itself. Aerodynamic considerations also confound the issue. Air-flow patterns around structures are complex and play a role in the variability of deposition fluxes. While knowledge is more developed for solids with smooth surfaces, real buildings have openings and overhangs that increase roughness (NAPAP 1987). The process and variability of precipitation interception by buildings is the least understood of factors to be considered in assessing materials damage (NAPAP 1987).

Table 6.1. Air pollution damage to materials

Materials	Type of Impact	Principal Air Pollutants	Other Environmental Factors	Methods of Measurement	Mitigation Measures
Metals	Corrosion, tarnishing	Sulfur oxides, other acid gases	Moisture, air, salt, particulate matter	Weight loss after removal of corrosion products, reduced physical strength, change in surface characteristics	Surface plating or coating, replacement with corrosion-resistant material, removal to controlled environment
Building stone	Surface erosion, soiling, black crust formation	Sulfur oxides, other acid gases	Mechanical erosion, particulate matter, moisture, temperature fluctuations, salt, vibration, CO_2, microorganisms	Weight loss of sample, surface reflectivity, measurement of dimensional changes, chemical analysis	Cleaning, impregnation with resins, removal to controlled environment
Ceramics, glass	Surface erosion, surface crust formation	Acid gases, especially fluoride-containing	Moisture	Loss in surface reflectivity and light transmission, change in thickness, chemical analysis	Protective coatings, replacement with more resistant material, removal to controlled atmosphere
Paints, organic coatings	Surface erosion, discoloration, soiling	Sulfur oxides, hydrogen sulfide, ozone	Moisture, sunlight, particulate matter, mechanical erosion, microorganisms	Weight loss of exposed painted panels, surface reflectivity, thickness loss	Repainting, replacement with more resistant material
Paper	Embrittlement, discoloration	Sulfur oxides	Moisture, physical wear, acidic materials introduced in manufacture	Decreased folding endurance, pH change, molecular weight measurement, tensile strength	Synthetic coatings, storing in controlled environment, deacidification, encapsulation, impregnation with organic polymers

Table 6.1. Air pollution damage to materials *(continued)*

Materials	Type of Impact	Principal Air Pollutants	Other Environmental Factors	Methods of Measurement	Mitigation Measures
Photographic materials	Microblemishes	Sulfur oxides	Particulate matter, moisture	Visual and microscopic examination	Removal to controlled atmosphere
Textiles	Reduced tensile strength, soiling	Sulfur and nitrogen oxides	Particulate matter, moisture, light, physical wear, washing	Reduced tensile strength, chemical analysis (e.g., molecular weight) surface reflectivity	Replacement, use of substitute materials, impregnation with polymers
Textile Dyes	Fading, color change	Nitrogen oxides, ozone	Light, temperature	Reflectance and color value measurements	Replacements, use of substitute materials, removal to controlled environment
Leather	Weakening, powdered surface	Sulfur oxides	Physical wear, residual acids introduced in manufacture	Loss in tensile strength, chemical analysis	Removal to controlled environment, consolidated with polymers, or replacement
Rubber	Cracking	Ozone	Sunlight, physical wear	Loss in elasticity and strength, measurement of crack frequency and depth	Add antioxidants to formulation, replacement with more resistant materials

Source: Linthurst (1984).

Damage to Metals

For metals, the most common damages are the loss of thickness due to corrosion and localized pitting or crevice corrosion (NAPAP 1987). Stress-assisted corrosion is less well understood. Dose-response coefficients for zinc corrosion as a function of SO_2 and time of wetness were given in Linthurst (1984) and MOI (1983). These coefficients appear to be the firmest quantitative parameters available for metals damage. Quantitative dose-response functions have been calculated for uncoated low-carbon steels and SO_2; however, it is unlikely that such steels would be unprotected in the environment in actual use. NAPAP (1987) concluded that much of the available research on dose-response functions for zinc and galvanized steel, plain carbon steel, pure copper, and pure aluminum is not useful for deriving quantitative dose-response functions for corrosion due to acidic deposition.

Damage to Stone

For masonry products, information on mechanisms and important factors is available but quantitative assessments are lacking. Cheng and Castillo (1984) attribute structural weakening of marble (by chemical conversion to gypsum) for the historic site of City Hall, Schenectady, N.Y., to sulfur compounds but offer no quantitative measure of the physical damages.

NAPAP (1987) notes that carbonate stone is particularly sensitive to dissolution of calcium carbonate by the hydrogen ion. Sulfur dioxide is thought to be important both because it contributes to rainfall acidity and because it is a reactant in the formation of gypsum. However, NAPAP (1987) also notes one research effort claiming that SO_2 was clearly not the major factor determining stone weight loss suggesting that all the variables affecting weathering must be considered in combination. Silicates are less sensitive to dissolution by acidic compounds than are carbonates (MOI 1983).

Damage to Paint/Coating Systems

According to NAPAP (1987), exterior coatings are the most complex materials being assessed for damage due to acidic precipitation. Furthermore, there is little understanding of how acidic com-

6: Materials Damage

pounds affect the physical/chemical processes that can impact paint and the substrate. MOI (1983) notes research indicating that SO_2 can penetrate paint film and in recent studies, high erosion rates observed for oil-base house paints were related to the loss of calcium carbonate, which is an extender pigment in paint. However, Haynie and Spence (1984) found that sulfur compounds had no impact on either oil or latex paint. Impacts upon the latex paint were found to occur as a result of NO_2 causing weight gain. NAPAP (1987) asserts that the understanding of pollutant impacts on surface coatings and the consequent development of dose-response functions is in its infancy.

Spence et al. (1975) determined dose-response relationships for erosion as a function of SO_2 and relative humidity for an oil-base paint and a vinyl-coil coating. In each case while SO_2 was a significant explanatory factor, it was dominated by relative humidity. MOI (1983) noted that since these dose-response relationships were determined under controlled conditions of simulated sunlight and high temperatures, they should not be used in ambient conditions.

Materials Inventory

Assessment of actual materials damage in quantitative terms requires an inventory of sensitive materials in the impacted regions. The potential resources at risk tend to exist in regions with high population density and high industrial activity rather than the sensitive ecological regions identified for aquatic and terrestrial impacts. The major industrial and population centers in both the eastern United States and eastern Canada receive high levels of acid deposition.

MOI (1983) concluded that without further surveys it was not possible to provide an inventory of renewable materials by sulfur deposition regimes. It did provide inventories of historic places at risk in both Canada and the United States due to sulfate deposition (in Canada) or SO_2 concentration (in the United States). However, these inventories merely recorded the number of sites or structures. In the absence of information concerning the type and amount of materials, these inventories are of little value in impact assessment.

NAPAP (1987) suggests that estimating material surface density comprises three steps. First, calculate the proportion of material i per building in class k, P_{ik}. Second, calculate the average surface area per building in class k, A_k. Third, calculate the number of buildings in class k per unit of land, N_k. The surface density of material i in class k, M_{ik} is then given by:

$$M_{ik} = P_{ik} \cdot A_k \cdot N_k$$

The total density for material i is then determined by totaling all k classes.

Work is proceeding under NAPAP to develop an inventory of construction materials for the northeastern United States. An inventory has been assembled for four metropolitan areas: Cincinnati, Ohio; New Haven, Connecticut; Pittsburgh, Pennsylvania; and Portland, Maine. Five categories of materials were employed: paint, stone, mortar, galvanized steel, and "other." The inventory did not include non-building structures such as bridges, storage tanks, and towers, which should be included in future efforts (Daum et al. 1987).

For cultural and historical materials, NAPAP argues that these must be treated as discrete resources that are not suited to the kind of modelling exercises used above for renewable construction materials.

Economic Impacts

Concern for the economic impacts has arisen more naturally in the materials receptor category than it has in the ecosystem research. However, there has not always been agreement on the appropriate approach. A pessimistic view of economic assessments of materials damage is sounded by Lee et al. (1985), who note that "economic studies performed by economists have been criticized by engineers and the engineering cost studies have been criticized by economists." In an optimistic and constructive vein, they suggest that "a joint program, where the economists tell engineers and chemists what data they need and the engineers/chemists provide the data, would help in estimating the costs of materials damage."

Crocker and Regens (1985) estimate $2.0 billion of materials damage is due to acid deposition in the eastern one-third of the United States. This is the highest damage category in their view. They note that the unit value of buildings is high and buildings constitute a large fraction of the nation's created wealth. Furthermore, for existing buildings there are fewer options available to adapt to the assault by acidifying compounds. The estimate was obtained by extrapolating the results of a 1970 Swedish household study to the United States. The Swedish study determined damages of $4.3 per person per year. Crocker (1985), in reviewing the $2.0-billion figure, noted that losses due to SO_2-induced paint erosion in Boston, Massachusetts, cost between $9.20 and $20.30 per person. As a result Crocker (1985) argued that the $2.0 billion is likely to be conservative. He also noted that the $2.0-billion figure did not include the additional impacts on non-metallic building materials such as stone, losses in unique private and public structures, cultural structures, or the manufacturing sector, which would imply even larger damages due to acid deposition. Based upon 1979 values of stocks and structures in the U.S. manufacturing sector and 1979 investment levels, a reduction in acid deposition that would extend replacement life by 6 months would be valued at $5.7 billion (Crocker and Cummings 1985).

Estimation of economic damages to materials is one goal of the NAPAP; however, NAPAP (1987) presents only a discussion of relevant methodologies. While no economic estimates are presented in NAPAP (1987), a discussion of some economic estimates made under the NAPAP is provided by Horst et al. (1987). The estimates are based upon the assumption that reducing acid deposition from current levels to natural background levels would increase the repair or replacement time. The resulting cost saving then is an approximate measure of the willingness to pay to reduce acid rain.

Horst et al. (1987) considered four building classes (single- and multiple-family residences, commercial/industrial buildings, and tax-exempt buildings), four materials (paint, zinc [galvanized steel], mortar, and stone), and seven uses of materials (walls, roofs, chimneys, gutters, downspouts, fencing, and window trim). The materials inventory was taken from the four cities mentioned above. Table 6.2 gives the aggregate estimates provided by Horst et al. (1987) for a 17-state region in the northeastern United States. Their paper also gives estimates by region.

Table 6.2. Acid deposition damage estimates to common construction materials in 17 states in northeastern United States

Category	Damages ($ million, 1984)
Paint	1179.0
Residential	906.8
Attributed to local SO_2	473.5
Total damages across all categories	2255.8

Source: Modified from Horst et al. (1987).

The table shows that over half of the estimated damages is attributed to painted surfaces, making them the most economically significant of the materials category (Wyzga 1987). Local SO_2 contributed only 20–25% of the total estimated damages. Residential structures accounted for almost half of the estimated damages. The region with the largest attributable damage consisted of the mid-Atlantic states, which suffer damages of $1151 million, or roughly half of the total and almost twice the next largest damage region. The eastern south central region had the least damage, with $29.8 million, or roughly 1% of the total damage. Horst et al. (1987) suggest that in light of uncertainties in the data, plausible damage estimates range from $688 million to $6.7 billion. The $2.3-billion estimate in Table 6.2 is very close to the Crocker and Regens (1985) figure, if no allowance is made for inflation.

These economic estimates have been criticized by Bradley (1987) and Glauthier and Mayer (1987). Bradley notes that the economic assessments focus on only one option available to building owners for response to materials degradation, which is to restore the structure to its original condition. Other possible responses not considered include (1) do nothing and suffer a reduction in service flow, (2) substitute less sensitive materials, and (3) relocate to a new building less susceptible to damage. Since the least-cost option may not be the one selected by the analysts, the damages may be exaggerated. Bradley is also concerned about the uncertainty of the relative causes of material damages by various pollutants. In order to increase confidence in economic estimates of materials damage, Bradley advocates the use of surveys to determine actual maintenance practices of owners and to estimate actual damage values.

Glauthier and Mayer (1987) use a risk-assessment methodology to evaluate the NAPAP approach to estimating materials damage. The steps in this methodology consist of (1) identifying the hazard, (2) assessing dose-response relationships, (3) assessing exposure, and (4) characterizing the risk.

Identifying the hazard requires determining hazardous substances and receptors at risk. Glauthier and Mayer question why NAPAP paid so much attention to mortar and so little attention to concrete, paving materials, roofing materials, and car paint. With respect to the dose-response relationship, Glauthier and Mayer note that research on paint concentrates on mass loss as the measure of damage rather than other forms of damage such as peeling or rusting, which were not evaluated. The mortar that was used in the dose-response was a simple lime mortar that has had little use in the United States during this century!

In assessing exposure they note the considerable uncertainty associated with extrapolating data from 4 cities to 117 metropolitan areas in the 17-state region. For risk assessment, they maintain that the NAPAP estimates ignore the fact that in most cases owners would undertake routine maintenance prior to the time that critical damage due to acid deposition would occur. This routine maintenance would be for reasons other than pollution-induced damages. For the mortar example, they note that it would take 229 years to reach the critical damage level! Only carbonate paints seemed to be close to the thresholds for which maintenance would be necessary.

Livingstone (1987) notes that the NAPAP paint damage function is based upon studies that pre-date the acid rain research program. He also questions the use of mass loss as the measure of damage when paint film may also be damaged by bloom (a whitish discoloration of oil-based films), cracking (brittle failure of the paint layer), and chalking (gradual loss of paint through destruction of the paint layer apart from carbonate mass loss).

Callaway et al. (1986) offer preliminary estimates of annual, increased maintenance costs of acidic deposition and precursor impacts upon cultural materials and transmission towers. For the cultural statues and monuments subgroup, they offer a range of $5–7 million ($ 1984); for the historical buildings they offer a wide range of $17–100 million. Transmission towers suffer increased maintenance costs of $2 million per year. They offer very little information about the basis for these estimates. The range of values for the cultural materials is based upon different discount rates (5% versus 10%). No reason is offered for the firm estimate for transmission towers.

At a special Air Pollution Control Association Conference session on the NAPAP interim assessment, the consensus of panelists and audience participants was that "data are insufficient to quantify important potential effects of acid precipitation on materials" (Lefohn and Krupa 1988).

References Cited

Bradley, R.A. 1987. Materials damage, acid deposition and the NAPAP assessment. J. Air Pollut. Control Assoc. 37:683.

Callaway, J.M., R.F. Darwin, and R.J. Nesse. 1986. Economic valuation of acidic deposition damages: Preliminary estimates from the 1985 NAPAP assessment. Water, Air, Soil Pollut. 31:1019–34.

Cheng, R.J., and R. Castillo. 1984. A study of marble deterioration at City Hall, Schenectady, N.Y. J. Air Pollut. Control Assoc. 34:15–19.

Crocker, T.D. 1985. Acid deposition control benefits as problematic. J. Energy Law Policy 6:339–56.

Crocker, T.D., and J.L. Regens. 1985. Acid deposition control. Environ. Sci. Tech. 19:112–16.

Crocker, T.D., and R.G. Cummings. 1985. On valuing acid deposition–induced materials damages: A methodological inquiry. In Acid Deposition: Environmental Economic and Policy Issues. Ed. D.D. Adams and W.P. Page, pp. 359–84. New York, London: Plenum Press.

Daum, M.L., F.W. Lipfert, and N.L. Oden. 1987. The distribution of materials in place. J. Air Pollut. Control Assoc. 37:681–82.

Franey, J.P., and J.E. Graedel. 1985. Corrosive effects of mixtures of pollutants. J. Air Pollut. Control Assoc. 35:644–48.

Glauthier, T.J., and F. Mayer. 1987. A critical perspective on NAPAP's methodology for materials damage assessment. J. Air Pollut. Control Assoc. 37:683–84.

Haynie, F.H., and J.W. Spence. 1984. Air pollution damage to exterior household paints. J. Air Pollut. Control Assoc. 34:941–44.

Horst, R.L., T.J. Lareau, and F.W. Lipfert. 1987. Economic evaluation of materials damage associated with acid deposition. J. Air Pollut. Control Assoc. 37:682–83.

Lee, S.D., J.M. Kawecki, and G.K. Tannahill. 1985. Evaluation of the scientific basis for ozone/oxidants standards. J. Air Pollut. Control Assoc. 35:1025–32.

Lefohn, A.S., and S.V. Krupa. 1988. Acidic precipitation: A technical amplification of NAPAP's findings. J. Air Pollut. Control Assoc. 38:766–76.

Linthurst, R.E., ed. 1984. The Acidic Deposition Phenomenon and Its Effects: Critical Assessment Review Papers. Vol. 2, Effects Sciences. EPA-600/8-83-016B, Washington, D.C.: U.S. EPA.

Livingstone, R.A. 1987. Critique of the carbonate mass loss model for paint damage functions. J. Air Pollut. Control Assoc. 37:684–85.

Memorandum of Intent on Transboundary Air Pollution (MOI). 1983. Final report of the impact assessment work group, Chap. 6. Ottawa and Washington, D.C.

National Acid Precipitation Assessment Program (NAPAP). 1987. Interim Assessment Report. Vol. 4, Effects of Acid Deposition. Chap. 9, Materials Damage. Washington, D.C.

Research Consultation Group (United States–Canada) (RCG). 1979. The LRTAP problem in North America: A preliminary overview. Ottawa and Washington, D.C.

———. 1980. Second report on the long range transport of air pollutants. Ottawa and Washington, D.C.

Spence, J.W., F.H. Haynie, and J.B. Upham. 1975. Effects of gaseous pollutants on paints: A chamber study. J. Paint Technol. 47:57–63.

Wyzga, R.E. 1987. Session chairman's summary and conclusions. J. Air Pollut. Control Assoc. 37:685–86.

CHAPTER 7

Human Health and Visibility Impacts

Health Impacts

Introduction

The consensus of opinion is that acid deposition is unlikely to have direct effects upon human health (MOI 1983). The discussion of health impacts in Linthurst (1984) was confined to the indirect effects on health only. However, it is acknowledged that precursor pollutants and ozone may have direct damaging effects on human health.

Direct Health Impacts

Ozone and nitrogen dioxide (NO_2) are deep-lung irritants that can produce pulmonary edema (a build-up of fluid in the lungs), in severe cases leading to possible death. In less severe cases, ozone exposure leads to substernal tightness, irritation of the mucous membranes, dry cough, and headache (MOI 1983). Those individuals who have existing respiratory ailments such as asthma, emphysema, or bronchitis are more sensitive to low-level exposures than healthy individuals. Ozone may also cause eye, nose, and throat irritations, while exposure to NO_2 increases one's susceptibility to infections (MOI 1983). Harrington and Krupnick (1985) found an inverse relationship between NO_2 concentrations and acute respiratory disease in Tennessee children; i.e., more illness was associated with lower pollution levels than higher ones.

Sulfur dioxide (SO_2) is also a respiratory irritant leading to broncho-constriction (a narrowing of the airway). The National Acid Precipitation Assessment Program (NAPAP 1987) reported evidence that acid sulfate aerosols can cause human effects. In a laboratory study of asthmatic and healthy volunteers, Avol et al. (1988) determined that the asthmatics' responses to both SO_2 and sulfuric acid (H_2SO_4) fog were roughly similar. However, the broncho-constriction was correlated more closely with SO_4 than with the H_2SO_4 fog and occurred to a greater extent in asthmatics than in the healthy subjects. They concluded that acid-fog exposure would lead to no pulmonary dysfunction and only slight respiratory symptoms, if any at all. In an earlier study of young exercising asthmatics exposed to H_2SO_4 aerosols, Avol et al. (1986) observed "no meaningful unfavorable responses even with moderately heavy exercise." Responses to SO_2 were greater than the responses to the H_2SO_4 aerosols. In their study of acute respiratory disease in Tennessee children, Harrington and Krupnick (1985) noted that the incidence of illness was positively correlated with the ambient sulfate level over the entire sample period. However, the relationship was not significant for reasons analyzed separately.

NAPAP (1987) notes that acidity may affect the toxicity of fine-particle aerosols and that titratable acidity, which includes undissociated weak acids, may be an even better determinant of the toxicity of aerosols. However, aerosol sulfate concentration is not a good predictor of this acidity.

Indirect Health Impacts

While there is little concern for the direct effects of acid deposition itself upon human health, there is concern that acid deposition could have indirect effects upon health. The major pathways for these indirect effects are through contamination of drinking water and the consumption of fish with increased body burdens of toxic heavy metals.

The issue of acid deposition and drinking water has been considered by McDonald (1985). As discussed in Chapter 5 on aquatic impacts, acidification of waterways can lead to the increased mobilization of toxic metals. Chief among these are mercury and aluminum.

The concern for mercury is more in its accumulation in fish than in affecting drinking water quality. However, 24% of rural U.S. households have shown contaminant levels in excess of the National Primary Drinking Water Regulations. Aluminum has been implicated in dialysis dementia, which is a central nervous system disorder. Aluminum in drinking water has not been related to acute or chronic diseases in humans (Linthurst 1984).

The major concern for drinking water quality comes from corrosion of household plumbing introducing contaminants into the water. The potential contaminants include copper, cadmium, lead, and asbestos. The presence of copper in drinking water is detected by taste, even at low levels. Thus it is likely that individuals will not knowingly continue to consume drinking water with dangerous copper concentrations. Consequently, the potential damage from acid-induced copper in drinking water is thought to be small.

Cadmium may be present in drinking water as a result of corrosion of galvanized pipe or of copper-zinc solder used in the plumbing system. The half-life of cadmium is 20 years; hence, it is viewed as a concern. Cadmium intake causes renal tubular damage.

Roughly 16% of a sample of 2654 rural U.S. households had drinking water with lead levels exceeding the maximum contaminant level (MCL) of 50 μg/L specified in the National Primary Drinking Water Regulations (Francis et al. 1984). Infants and fetuses are particularly sensitive to lead contamination, the effects of which include mental retardation for infants. IQ deficiencies and behavioral problems can result even at low levels of lead in the blood.

While it is known that airborne asbestos is carcinogenic, the results of increased ingestion of asbestos-contaminated water is not yet fully known.

Many of the problems with corrosion in pipes are reduced by allowing the pipes to flush for several minutes before obtaining drinking water. Copper levels, for example, increase as water stands overnight in pipes and taps but rapidly decline when flushing takes place. Concentrations of lead in corrosive drinking-water systems are three times higher in first-flush samples than after flushing takes place.

Groundwater is not thought to be at risk from acid deposition except in poorly buffered areas. However, drinking water drawn from surface waters undergoing acidification may pose problems. The greatest concern appears to be for those individuals drawing water from cisterns and small, private or rural facilities. Large managed facilities should adjust pH to appropriate levels to lower corrosiveness and to avoid health problems either directly or from leaching (NAPAP 1987). McDonald (1985), however, notes that 68% of large suppliers in the United States do distribute moderately aggressive (i.e., corrosive) water, and 84–92% of drinking waters in the northeastern United States were very aggressive.

Another main concern for indirect health effects of acid deposition is the consumption of edible fish from acidified waters. As noted, there is evidence of elevated mercury body burdens in fish from poorly buffered lakes. One hypothesis is that lakes with low pH produce larger amounts of monomethyl mercury, which is then taken up by aquatic biota. Increased foraging by predator fish increases the bioaccumulation of the mercury in the tissues of fish. However, the correlation between mono- and dimethyl mercury content and pH was poor (MOI 1983).

The bioaccumulation of mercury is greater in the larger piscivorous fish (which are among the more prized sport fish and make up the majority of the eaten, yearly catch) than in the planktivores (MOI 1983). Freshwater pike and trout showed the highest average concentrations (Linthurst 1984). The effects of methyl mercury result in damage to the central nervous system. Symptoms include loss of sensation, impaired hearing, and constriction of visual fields (Linthurst 1984).

The maximum safe intake of mercury is 30 μg per day. For an average consumer, the concentrations in pike and trout would lead to daily intakes of 10.4 μg from pike exclusively and 5.2 μg from trout exclusively, which are well below the maximum safe intake level (Linthurst 1984). Populations, such as Native American populations, that are more dependent upon fish from acidified waters than average consumers may need to be concerned. Some mild neurological abnormalities in adult Cree men have been noted as has an association between Cree boys' reflexes and the mercury level of their mother's blood during pregnancy (MOI 1983).

Those individuals who consume larger-than-average quantities of water with low concentrations of mercury are more at risk than those consuming an assumed average of 2 L per day. For diabetics and individuals with certain kidney ailments who consume more than average, and for those who consume more during hot weather or during periods of physical exertion, intake levels may be as high as 10

L per day. Based on body weight, the fluid consumption for infants and children is also relatively high compared to adult consumption (McDonald 1985).

Visibility Impacts

There seems to be general agreement that acid deposition contributes to a deterioration in visibility in the northeastern portion of North America. Hare (1986) claims that reduced visibility throughout the midwestern and northeastern North America is one of the major impacts of SO_2 emissions. He further asserts that sulfate particle scattering causes a blue sky to be replaced by a white sky.

In addition to the sulfate particles, other compounds may affect visibility. Nitrogen dioxide and particles may produce brownish plumes (MOI 1983), and when NO_2 absorbs blue light, it produces a yellow hue (NAPAP 1987). Airborne particles may produce black, gray, or bluish plumes (MOI 1983). Nitrates are generally considered of little consequence to visibility in eastern North America (MOI 1983). Research shows that sulfate accounts for 94% of the variability in light scattering in the Allegheny Mountains of Pennsylvania (NAPAP 1987). Visibility impairment was not an effect considered in the critical review papers on acid deposition prepared by Linthurst (1984).

Mathai and Tombach (1987) observe that visibility in nonurban areas of the eastern United States decreased between the mid-1950s and mid-1970s with slight improvement occurring since the mid-1970s. Meteorology was found to play a significant role in these trends. Furthermore, sulfur accounted for 46% of annual fine mass and 60% of mean summer fine mass, assuming that all sulfur existed as ammonium sulfate.

Economic Valuation of Health and Visibility Impacts

Valuation of Health Impacts

There has been very little economic valuation of health damages due to acid deposition. NAPAP (1987) did not address the economics of health impacts at all. As discussed above there is little concern for

7: Human Health and Visibility Impacts

direct health damage as a result of acid deposition.

There remains, however, concern for potential indirect health effects attributable to acid deposition. NAPAP (1987) notes that drinking water contaminated with copper will have a taste that discourages ingestion and thus concludes copper contamination "is not considered a problem." The health problem is avoided because people will not drink the contaminated water because the taste alerts them; however, the economic issue remains because the source of the contamination must then either be replaced (i.e., new plumbing) or an alternative water supply must be found. The cost associated with this adjustment is a measure of the value of "health damage" avoided and needs to be considered. Estimates of these adjustment costs are elusive at present.

In managed community drinking water systems, proper management adjusts pH to appropriate levels so that contaminants should not be elevated due to acidity. The economic impact in this case is the management cost of additional liming procedures attributable to acidity from atmospheric sources. To the extent that these procedures are not fully carried out, the economic damage will exceed the additional management costs.

The economic impacts associated with contaminated fish may be minimized by either limiting the intake of the fish likely to exhibit elevated body burdens or by substituting other food sources for these fish species. The economic impact of such substitution is the loss in satisfaction in having to move to a possibly less-preferred alternative. No data exist on this to the author's knowledge.

While NAPAP (1987) concluded that no estimate of the human population at risk from elevated mercury content in fish is available, Linthurst (1984) did mention one relevant study. A survey of 25,000 individuals, representative of the U.S. population, revealed that 47 individuals exceeded the maximum safe daily intake by a small margin and 23 of these were mainly consumers of freshwater fish. Thus, the risk of being exposed to excess mercury from freshwater fish is roughly 1 in 1000.

Valuation of Visibility Impacts

Reduction in visibility may affect aesthetic enjoyment and transportation safety. The reduction in aesthetic enjoyment in residential areas may be expressed in reduced household property value (other things being equal). In the case of recreation, economic losses due to visibility reduction may not be reflected adequately by market indicators. Tourism and recreation by non-local residents may be lower as a result of poorer visibility than competitor regions. Loucks (1987) argued that recreation and tourism in the state of Indiana suffer as a result of the poor visibility conditions caused by acid deposition.

The superb visibility conditions that exist in the American Southwest and the Rocky Mountain region are an important aesthetic factor in the enjoyment of the recreation experience for tourists as well as a benefit for local residents. Several economic studies have been undertaken since the mid-1970s to determine the value that households place on visibility. These studies were instrumental in developing the techniques of analysis that are now included under the generic term of "contingent valuation methods" for valuing non-marketed commodities. These methods rely on direct surveys, which may be face-to-face interviews, telephone interviews, or mail surveys. Respondents are asked to value specified changes in some environmental attribute. Several of the earlier studies are reviewed and compared by Schulze et al. (1981).

Randall et al. (1974) used a bidding-game framework to assess the value of reducing aesthetic environmental damage associated with the Four Corners power plant and the Navajo mine near Fruitland, New Mexico. Respondents were shown three sets of photographs depicting varying degrees of aesthetic impairment.

The base case, scenario A, showed the most severe damage, spoil banks associated with strip mining, prior to 1970 and prior to the installation of additional emissions-control devices and with electricity transmission lines interrupting visual impression. In scenario B, the plant is operating with additional emission controls and hence improved visibility; transmission lines are less obtrusive and spoil banks are leveled but not revegetated. Scenario B represents an intermediate level of damage, which corresponded roughly with the situation at the time of the survey. The final scenario, C, showed the plant with no visible emissions and with minimal environmental damage. Another photograph showed arid land in its natural state with no evidence of transmission lines or spoil from strip mining.

Given these scenarios, individuals were asked how much their households would be willing to pay to obtain scenario B rather than A; C rather than A. Using a sales tax as the payment mechanism, the average estimate of willingness to pay was $50 per year to obtain scenario B and $85 per year to obtain scenario C rather than A. In general, these estimates overstate the value placed upon visibility alone, since the respondents were valuing a composite environmental commodity impacted not only by visibility impairment due to particulate emissions but also the aesthetic insults of transmission lines and mine spoil banks.

Brookshire et al. (1976) used a bidding-game approach to estimate the possible aesthetic damage associated with the proposed construction of the Kaiparowitz power plant near Lake Powell in the Glen Canyon National Recreation Area. At the time of the study, the 2400-MW Navajo power plant had been constructed and was scheduled to become fully operative. As in the Randall et al. (1974) study, three scenarios were depicted by photographs. Scenario A showed a possible site but with no power plant having been built. This represents a pristine base case. Scenario B showed a power plant existing but no impairment of visibility from emissions, and C showed a power plant with an emissions plume impairing visibility.

The bidding-game approach was used to determine the largest amount that individuals would be willing to pay to prevent scenarios B or C from occurring. The average daily bid in the form of an additional entrance fee was $1.58 to prevent a deterioration from A to B, and $2.77 to prevent a deterioration from A to C. Brookshire et al. (1976) do not supply individual, annual willingness-to-pay figures. This would require knowledge of annual visitation. The value reported for avoiding scenario B is the value for avoiding the visual insult of the power plant as a physical structure marring the landscape. The value for avoiding scenario C includes both this value and the value for avoiding the visibility reduction. This amounts to an additional $1.19 per visitor day. The study by Rowe et al. (1980) was designed to focus on the visibility issue for the Four Corners region per se, also using the bidding-game framework. The respondents were asked to value potential decrements in visibility range as represented by photographs. The visibility range changes were from 75 miles to 50 miles to 25 miles. The average annual bid to prevent the deterioration in visibility from 75 miles to 50 miles was $57, while the bid to prevent deterioration to 25 miles was $82.20 per year.

Schulze et al. (1981) suggested that the Rowe et al. (1980) values are comparable to the Randall et al. (1974) values if allowance is made for price inflation between study time periods and the fact that Randall et al. (1974) considered other adverse impacts in addition to visibility range reduction. Randall et al. (1974) determined what are called "compensating surplus" measures of the economic welfare change, while the measures quoted above for comparison by Rowe et al. (1980) are "equivalent surplus" measures. The discrepancy between these types of measures has been the subject of considerable debate in the economics literature. Curiously, Rowe et al. (1980) claimed that Randall et al. (1974) "did not report compensating measures." This may have resulted from the common notion of compensation "requiring a payment to the respondent." This is not the issue in the Randall et al. (1974) study.

The value of preserving visibility in the Grand Canyon and parklands of the American Southwest was estimated by Schulze et al. (1983). The monthly average values bid to preserve visibility in the Southwest including the Grand Canyon were $6.61 for Denver residents, $9.31 for Chicago residents, $8.23 for Albuquerque residents, and $9.64 for Los Angeles residents. These monthly estimates produce annual estimates of the value of preserving visibility that range from roughly $79.00 to roughly $110.00.

Melamed (1984) cites an unreleased study, which concludes that a 22% visibility improvement in the Ohio River Basin would be worth $32–185 annually to the typical household in Cincinnati. The annual figures in this study bracket those determined for the U.S. Southwest reported above. Extrapolating to the eastern United States, Melamed (1984) reports an aggregate benefit of between $1.3 and $10 billion per year. The "best" estimate was $4.0 billion per year. This unreleased study was sponsored by the Electric Power Research Institute.

References Cited

Avol, E.L., K.R. Anderson, W.S. Linn, D.A.S. Shamoo, J.D. Whynot, and J.D. Hackney. 1986. Respiratory responses of exercising asthmatic volunteers exposed to sulfuric acid aerosol. J. Air Pollut. Control Assoc. 36:1323–28.

———. 1988. Short-term respiratory effects of sulfuric acid in fog: A laboratory study of healthy and asthmatic volunteers. J. Air Pollut. Control Assoc. 258–63.

Brookshire, D.S., B.C. Ives, and W.D. Schulze. 1976. The valuation of aesthetic preferences. J. Environ. Econ. Manage. 3:325–46.

Francis, J.D., B.L. Brower, W.F. Graham, O.W. Larson, III, J.L. McCaull, and H.M. Vigorita. 1984. National statistical assessment of rural water conditions, Vol. 3. Cornell University for the Office of Drinking Water, Environmental Protection Agency. Washington, D.C.

Hare, K.F. 1986. Air quality and the energy-environment interface. In Managing the Legacy. Minister of Supply and Services, Canada.

Harrington, W., and A.J. Krupnick. 1985. Short-term nitrogen dioxide exposure and acute respiratory disease in children. J. Air Pollut. Control Assoc. 35: 1061–67.

Linthurst, R.A., ed. 1984. The Acidic Deposition Phenomenon and its Effects: Critical Assessment Review Papers. Vol. 2, Effects Sciences, Chap. 6. EPA-600/8-83-016B. Washington, D.C.: U.S. EPA.

Loucks, O. 1987. The Ecology of acid rain. A presentation to the symposium, Acid Rain: The Ecology, the Economics and the Politics of a Problem, held at Indiana State Museum, Indianapolis.

McDonald M.E. 1985. Acid deposition and drinking water. Environ. Sci. Technol. 19:772–76.

Mathai, C.V., and I.H. Tombach. 1987. A critical assessment of atmospheric visibility and aerosol measurements in the eastern United States. J. Air Pollut. Control Assoc. 37:700–707.

Melamed, D. 1984. Washington report. J. Air Pollut. Control Assoc. 34:200.

Memorandum of Intent on Transboundary Air Pollution (MOI). 1983. Final report of the impact assessment work group. Ottawa and Washington, D.C.

National Acid Precipitation Assessment Program (NAPAP). 1987. Interim Assessment Report. Vol. 4, Effects of Acid Deposition. Chap. 10, Effects on Human Health and Visibility, Washington, D.C.

Randall, A., B. Ives, and C. Eastman. 1974. Bidding games for valuation of aesthetic environmental improvements. J. Environ Econ. Manage. 1:132–49.

Rowe, R.D., R.C. d'Arge, and D.S. Brookshire. 1980. An experiment on the economic value of visibility. J. Environ. Econ. Manage. 7:1–19.

Schulze, W.D., R.C. d'Arge, and D.S. Brookshire. 1981. Valuing environmental commodities: Some recent experiment. Land Econ. 57:151–71.

Schulze, W.D., D.S. Brookshire, E.G. Walther, K.K. MacFarland, M.A. Thayer, R.L. Whitworth, S. Ben-David, M. Malm, and J. Molenar. 1983. The economic benefits of preserving visibility in the national parklands of the southwest. Nat. Resour. J. 23:149–73.

CHAPTER 8

Control Options and Strategies

Options for Reducing Emissions from the Electric-Power Industry

Given a decision to reduce acid deposition, it is necessary to reduce the emissions of its precursors. In order to reduce these emissions it is necessary either to reduce the level of the economic activity that is generating them or to change the method by which that economic activity is conducted so as to reduce emissions per unit of output. Since the acid rain debate in the United States concentrated on the sulfur dioxide (SO_2) (and nitrogen oxides, NO_x) emission levels of electric-power plants in the upper Ohio Valley using high–sulfur content coals, most of the discussion in this chapter focuses on the electric-power industry.

Reduction of electricity production by the electric-power industry in the Ohio Valley would necessitate either a reduction in electricity consumption by the relevant consuming groups, or the import of electrical power from outside the region to supply the shortfall caused by a reduction in production. Indeed, some prominent politicians from major coal-producing regions have suggested that the Canadians are pushing acid rain controls so as to expand their electricity export markets.

Changing the method by which the electricity is produced affords a wide range of alternatives. These include changing the fuel used, improving the quality of the coal currently used, improving in-process technology to reduce emissions, and using devices to capture

8: Control Options and Strategies

residuals thereby prohibiting their release to the atmosphere.

Changing fuels actually presents a wide variety of potential options. Some of these are most relevant for firms that would phase out existing coal-fired plants and replace them with new facilities. An obvious choice in this area is to move to non-fossil fuels such as nuclear power. This decision will be influenced by the usual economic considerations such as relative capital costs of the facility and fuel costs, and the acid deposition reduction requirement, but also by the anti-nuclear movement. From a societal point of view this switch represents a trade-off between acid rain precursors (and the greenhouse-effect precursors) and nuclear-waste disposal (and the China Syndrome concerns).

Electric-power generation could also use solar, wind, or hydro resources. A major concern here, however, is the availability of the relevant resource in the relevant region. Since the upper Ohio Valley is not as well suited to solar and wind electric-power generation as the American West and Southwest, these are unlikely to be realistic options. The hydro option may be available. All of these are relevant to decisions concerning new plant construction.

Remaining with fossil fuels still provides several choices, most of which can be accomplished by converting existing plants. These include the use of oil and natural gas rather than coal. The switch from high-sulfur coal to low-sulfur coal provides an obvious fuel-switching option that allows for emission reduction. This possibility has generated the most "heated" reaction among the high-sulfur coal-producing states as will be seen later.

Remaining with current coal usage still has a number of options available to reduce emissions to the atmosphere. The first option consists of coal-cleaning procedures designed to remove sulfur from the otherwise high-sulfur coal, thus improving the quality of the fuel from an environmental point of view. The most basic method of cleaning consists of "physical" coal cleaning, which removes up to half the sulfur in some coals. Physical coal cleaning relies on different densities between coal and its impurities for separation to take place.

Experimental work is suggesting that the use of biological agents to cause pyrite particles to sink more rapidly may double sulfur removal (Shepard 1988). Physical and biological coal-cleaning processors do not remove organic sulfur and hence these procedures are not thought to be sufficient to meet stringent sulfur-emissions standards. Since organic sulfur is chemically bound into the coal molecule, it cannot be separated by physical separation procedures. There are chemical separation techniques in the development stage that may be useful complements to the physical coal-cleaning processes (Shepard 1988).

A further option for existing coal-fired plants is to pursue retrofitting with emission-control technology. The most basic technology in this category is flue gas desulfurization (FGD), known popularly as "scrubbing." Sulfur is removed from flue gases using an alkaline slurry, which produces a large amount of sludge that requires disposal. The annual sludge from a 1000-MW plant would cover a square-mile area to a depth of 1 foot (Shepard 1988). This leads to a consideration of possible commercial products from scrubber sludge.

An alternative process to wet scrubbing is termed "spray drying." With this technique a slurry, such as calcium hydroxide, dries as it reacts with sulfur dioxide in the exhaust stream (Shepard 1988). The by-products of this reaction are captured in electrostatic precipitators or fabric filters, or fallout in the spray tower (Shepard 1988). This process is currently being used commercially in low-sulfur coal-fired power plants; work is in progress to determine possible application to high-sulfur coal-fired plants (Shepard 1988).

Another set of dry desulfurization technologies is called "dry sorbent emission control" technologies (Muzio and Offen 1987; Shepard 1988), in which a dry powder is injected into the furnace or postfurnace area. Pilot studies have shown that downstream injection is more effective than injection to the burner itself (Hendriks and Nolan 1986). This is related to the optimum temperature range of 850–1250°C for good sorbent reaction and a high degree of sulfation. Calcium hydroxide performed better as a sorbent than lime or raw limestone (Hendriks and Nolan 1986). These dry calcium sorbent injection processes appear to have lower SO_2 removal efficiencies than wet scrubbing (Muzio and Offen 1987). Wet FGD processes remove 90% of the sulfur, while the dry sorbent processes remove 40–60% (Muzio and Offen 1987).

Dry injection of lime downstream from an electrostatic precipitator (or other particulate device) and upstream from a baghouse is also being studied (Shepard 1988). Injection of sodium compounds upstream from the electrostatic precipitator is the focus of another pilot project (Muzio and Offen 1987). The relevant sodium sorbents are sodium bicarbonate and sodium sesquicarbonate, which can be the naturally occurring minerals nahcolite and trona or chemically

8: Control Options and Strategies

processed derivatives. The sodium-based injection also removes about 10% of the NO_x but results in stack-plume discoloration due to the production of NO_2 (Muzio and Offen 1987).

The sodium-injected process has succeeded in removing 70–90% of the SO_2 when used with a fabric filter as opposed to 60% with an electrostatic precipitator (Muzio and Offen 1987). The sodium-injection process is considered to be commercially available, while the calcium-injection processes are more recent discoveries (Muzio and Offen 1987).

The various dry-injection processes may solve the solid-waste disposal problems associated with the wet FGD processes. For example, nahcolite refuse can be buried in the same pits as the fly ash from the burned coal provided the pit is lined to prevent soluble sodium from entering the groundwater system (Purdy 1988).

Work is also proceeding for retrofitting existing facilities for NO_x controls. The two main approaches include modifying the combustion process to reduce NO_x production and removal of the produced nitrogen compounds from the flue gases (Shepard 1988). Changing the combustion process has a cost advantage (Maulbetsch et al. 1986; Shepard 1988). Reduced NO_x formation can be achieved by reducing oxygen levels and lowering flame temperatures (Maulbetsch et al. 1986; Shepard 1988). Various methods for achieving these modifications are discussed by Maulbetsch et al. (1986). These combustion-process changes are the only NO_x controls that have been used in the United States, and most of these have been for new power plants (Maulbetsch et al. 1986). The use of low-NO_x burners, perhaps combined with "overfire air," is the preferred method for most boiler designs other than cyclone boilers for which reburning is the option (Offen et al. 1987).

Controlling the NO_x after combustion may be achieved using selective catalytic reduction (SCR) — a technique recently developed in Japan. Under ideal conditions this has achieved 80% removal. This approach is considerably more expensive than the low-NO_x burner approach (Maulbetsch et al. 1986). SCR is being used in Germany and Austria and is achieving 65–80% removal of NO_x. It is not clear that the European experience will be a good guide for U.S. applications because of differences in coal (Offen et al. 1987). Most applications of SCR have used low-sulfur coal, so there is uncertainty regarding the effectiveness of the process with medium- and high-sulfur coals (Shepard 1988).

Electric Power Research Institute (EPRI) research on clean coal technologies for new plants concerns improved pulverized coal (IPC), fluidized-bed combustion (FBC), and gasification combined cycle (GCC) (Shepard 1988). The IPC approach uses materials and design modifications based upon existing technology but designed to increase the heat rate by 15–20% (Shepard 1988). FBC technology may be either atmospheric or pressurized. Atmospheric FBC is more traditional than pressurized. Atmospheric FBC adds direct in-boiler SO_2 control to a typical coal-fired steam boiler. These plants can burn any coal as well as other fuels (Moore 1987). Pressurized FBC extends the atmospheric technology to more compact, modular designs with higher efficiency (Moore 1987). This technology has several unresolved technical problems, however. Shepard (1988) suggests that the greatest advantages of FBC technologies are the fuel flexibility and the control of pollutants in the furnace. The removal of sulfur is up to 95% without using a scrubber, and because of the relatively low operating temperature, the NO_x emissions are kept low (Shepard 1988). FBC may also be used as a retrofit technology.

The GCC is designed to produce a clean-burning fuel by gasifying the coal and separating the pollutants (Shepard 1988). A combustion turbine burns the clean fuel while a steam turbine is powered by the waste heat from the gasifier and the combustion turbine (Shepard 1988). Like FBC technology, GCC can be built-in modules. A GCC plant operating in California burning high-sulfur coal is still able to meet stringent emissions standards for SO_2, NO_x, and particulates; the solid waste produced has been judged nonhazardous (Shepard 1988).

In terms of meeting new source performance standards, the various technologies ranked from lowest potential to highest are pressurized FBC, baghouse-equipped atmospheric FBC, IPC with FGD, GCC, and a fuel-cell plant coupled with a GCC (Moore 1987).

Given the different pollutants that may be subject to control, it makes sense to pursue integrated environmental control (IEC) design procedures. IEC procedures coordinate the control of SO_2, NO_x, and particulates rather than attempting to control each in isolation of the other.

Research has considered the use of a baghouse or an electrostatic precipitator (ESP) for particulate control followed by a wet scrubber. Results indicate that better SO_2 removal occurred with the use of a baghouse than an ESP (Carr 1986). Unfortunately, according to Carr

(1986), the ESP–wet scrubber combination is more common. Baghouses are more efficient collectors of particulates (removing 99.9%) than ESPs. It seems that the particulates that escape the ESP hinder scrubber performance (Carr 1986).

Heat-improvement procedures for the power plant, such as the addition of low–excess air burners, could reduce the production of flue gases. This would permit the use of a smaller FGD system, for example, and less solid waste would be generated (Carr 1986; Shepard 1988).

High-temperature filtration methods suggest that the removal of SO_2, NO_x, and particulate matter may be combined into one step (Shepard 1988). Activated char and electron-beam irradiation are techniques that may collect SO_2 and NO_x simultaneously (Carr 1986; Shepard 1988).

Strategies for Implementing Emissions Reductions

The method selected for reducing emissions will be determined to a degree by the type of implementation scheme specified in the legislation. There are two alternative types of implementation schemes that can be used to bring about the reductions: one is based upon market incentives and allows the firm freedom of choice; the other is referred to as a command-and-control scheme, which considerably reduces the feasible set of actions or the flexibility of actions the firms may use in meeting reductions.

Command-and-Control Policies

In the command-and-control approach, the pollution-control legislation specifies not only the size of the emission reduction but may also stipulate the method by which that reduction must be achieved. Another form of command-and-control policy is one that requires the same reduction (or percentage reduction) from all sources. The command-and-control procedures are well illustrated by the 1977 amendments to the U.S. Clean Air Act, which set new standards for new sources of air pollution. The amended New Source Performance Standards (NSPS) require that firms building plants after the establishment of the new standards must reduce emissions consistent with the use of the best technological system (Gallogly

1981–1982). For fossil fuel–fired sources, the amendment further requires the achievement of a given percentage reduction in emissions from that which would be emitted in the absence of controls and independent of the actual original sulfur content of the fuel. The amendments also specify a policy of prevention of significant deterioration (PSD) of air quality in "clean" or pristine regions of the United States. Under the provisions of this policy, proposed facilities must use best available control technology (BACT). The 1988 International Nitrogen Oxides Protocol requires signatories to use best available technologies on new sources (*Coal Week* 1988).

The Clean Air Act amendments and the NO_x protocol eliminate the option to achieve similar emission reductions through other means such as the use of low-sulfur coal or coal cleaning. As a result of this reduced flexibility of command-and-control policies, economists tend to believe that these policies will be more costly than policies based upon market incentives.

Market-Incentive Policies

There is an extensive body of literature on the economic theory of market-based incentives for pollution control. Early work concentrated upon taxing the output of polluting firms. This leads to increased costs of production and the level of production will decrease as the product price rises, inducing a reduction in consumer demand. If emissions are directly related to the level of production, then this output tax will ultimately reduce emissions. In the present case this is equivalent to placing a tax upon electricity output, which drives up electricity prices, which encourages energy conservation. This conservation then leads to a drop in electricity production and hence less coal is burned and less emissions are produced. This approach ignores the technological options that may be available to reduce emissions without necessarily reducing electricity outputs. It also offers no incentive for firms to develop new technologies for reducing emissions.

Rather than taxing electricity output, authorities could place a tax upon those inputs that contribute the pollutants of concern. The firm thus has an incentive to economize on the use of the taxed input in order to reduce the tax burden. As the firm substitutes other inputs in the production process, the emissions will decline. In the present case a tax on coal would encourage either a reduction in coal usage

8: Control Options and Strategies

and hence emissions or perhaps a switch to alternative fuels, such as natural gas, oil, uranium, etc., not subject to the tax. In 1983 Congressman Les Aspin (D. Wis.) proposed a bill (H.R. 4483) with a sulfur-in-fuel tax of $.50 per pound of sulfur as a means to control acid rain. A differential tax on high- versus low-sulfur coal will encourage a shift away from high- to low-sulfur coals and reduce emissions. While the input tax provides for a more flexible response by the firms, it still precludes incentives for technological improvements in high-sulfur coal use that may be better than the alternatives currently available, unless credit is given for such removal. For example, the Aspin bill included a rebate equal to the tax rate for sulfur removed prior to release (Streets and Veselka 1987).

Since the intention of taxation is to reduce emissions, the emissions themselves should be the target of the taxation. A tax on emissions encourages the firms to reduce their emissions in the most economic fashion from the firms' point of view. Excluding any biases introduced by other government policies, this also produces a least-cost adjustment to the achieved emissions reduction from society's point of view. The impact of other government policies on the financial economics of alternative responses by firms will be discussed in Chapter 9.

An emissions tax was central to the Acid Rain Bill H.R. 2497 introduced by Congressmen Judd Gregg (R. N.H.) and Thomas Downey (D. N.Y.) in 1987. The tax scheme, termed a SANE tax, would place a sliding tax on *s*ulfur *a*nd *n*itrogen *e*missions. For SO_2 the range was $.15–.45 per pound and for NO_x the range was $.10–.45 per pound. As with other acid rain bills, this one was unsuccessful. More will be said about this later.

The emissions tax is criticized in practical applications for not being able to guarantee that emissions will in fact be reduced. It is feared that firms will simply pay the fee and continue emissions at the prior level (Regens and Rycroft 1986). A spokesperson for the National Association of Regulatory Commissioners levied this criticism at the Gregg-Downey SANE tax (*Coal Week* 1987).

It is true that an arbitrarily chosen tax may not achieve as large a reduction in emissions as might be desired. Firms will not reduce emissions if the cost of reducing emissions is more than the tax burden; however, this suggests that the tax rate is too low. Increasing the tax rate will increase the tax burden of any given emissions level. In principle, the authorities need only vary the tax rate until the appropriate adjustments are made by the firms. However, as noted by Montgomery (1972), this iterative method is cumbersome and politically unattractive. It creates considerable uncertainty for the firm in planning alternatives and is likely to produce a "wait-and-see-where-the-tax-settles" response by the firms.

Concern for the size of the resulting emissions achieved using the tax scheme leads to consideration of a dual market scheme in which the authorities control directly the quantity of emissions. A system such as this was first proposed for water pollution control by Dales (1968). In this market scheme the authorities issue a number of emission permits or emission licenses. The sum of the emission entitlements embodied in these permits becomes the amount of legally allowed emissions. The permits are then auctioned or allocated to firms by the authorities, with resale or transfer between firms being allowed. The permits will command a positive price, provided the total permit emission level is below the current emission level. In a simple market setting, the market price established for the permits will equal the emissions tax that would generate the same quantity of emissions as is controlled directly by the available permits. It is for this reason that the emission permit scheme (a quantity control) is said to be "dual" to the emission tax scheme (a price control).

An emission permit scheme (EPS) eliminates the concern over the overall emission level and allows firms the flexibility to decide upon the most economic response to controlling emissions. Again, in the absence of other biasing government policies, the emissions reduction is achieved in the least-cost fashion.

If there is a single, defined, homogeneous airshed, then the transferable EPS works reasonably well. Firms may trade permits with no change in total emissions and hence no change in air quality within the airshed. Montgomery (1972) noted the problems that arise for a transferable EPS system when there are multiple airsheds that may be impacted by any one firm as a result of air transport. Essentially, the problem is that trading permits among different firms may leave total emissions across all airsheds constant but may alter air quality within airsheds in a manner dictated by emitter-receptor transfer coefficients. Consider a simple example of two airsheds, each with one firm whose emissions stay entirely within its own airshed. A trade of permits between these two firms leaves total emissions constant but improves air quality in one airshed at the expense of the other. With multiple airsheds and many firms, each of which sends emissions to several airsheds, the outcome of a given trade is likely to be quite varied.

Montgomery (1972) proposed an alternative market scheme in which permits are rights to deposit pollution (pollution licenses) in specified receptors. Air quality is then controlled within each airshed by regulating the quantity of permits within the airshed. Krupnick et al. (1983) refer to this as an "ambient-based" permit system (APS). All firms must carry a portfolio of permits for all receptors receiving their emissions. Montgomery (1972) showed that 1) the initial allocation of fixed permits across firms could be arbitrary and 2) that this system achieves the least-cost solution for the predetermined air quality at each receptor. Montgomery (1972) also showed that it would be possible to redesign the EPS to allow for spatial aspects, but each permit must specify entitlement to emit according to the trading pair and the associated transfer coefficients. Furthermore, there were additional restrictions upon the initial allocations. This system would be an administrative *nightmare.*

Krupnick et al. (1983) argue that the Montgomery pollution license scheme would likely have high transactions cost as a result of the firms having to operate in several markets. Further, they prefer a modified version of Montgomery's EPS, although they concede that his original system is a nightmare for the administering agency attempting to achieve the least-cost solution.

Krupnick et al. (1983) suggest a hybrid formulation that combines features of Montgomery's APS pollution license with the EPS. In their scheme, permits are specified in terms of source emissions, but they are not necessarily traded on a one-for-one basis. The permits are not associated with any specific receptor. However, when a firm purchases permits either for a new source or an expanded one, it must also purchase sufficient permits from other emission sources to offset their additional contribution. Krupnick et al. call their system a pollution offset scheme (POS), which is designed to maintain regional airshed air quality as does Montgomery's APS. Their POS system achieves the least-cost solution and permits arbitrary initial allocation of the permits, unlike Montgomery's system.

McGartland and Oates (1985a) show that the Krupnick et al. (1983) POS is in fact mathematically identical to the Montgomery APS from the perspective of polluting firms operating with perfect competition. For non-competitive situations McGartland (1985a) argues that the Montgomery APS may indeed have high transactions costs while the Krupnick et al. POS allows firms to "free-ride" and to benefit from other firms' transactions.

Krupnick et al. (1983) suggest that the emissions trading policy authorized by the Environmental Protection Agency (EPA) facilitates the implementation of their POS proposal. This EPA policy included "offsets," "bubbles," and "emission reduction credits" (ERCs). The offset provision requires new sources entering a metropolitan area that has excessive air pollution concentration to find polluters who are phasing out their pollution. The new source then uses this reduction as an offset to its own pollution. Jones and Tybout (1986) suggest that cases exist whereby a new source has paid existing sources to reduce emissions. In one case cited a company was willing to "donate" an $18-million scrubber to another company in order to obtain an offset. The other company was willing to accept the deal but other problems caused the "new" source to withdraw its entry application. The 1986 Emissions Trading Policy Statement indicates that offsets may also be used in attainment areas to demonstrate protection of "prevention of significant deterioration" or visibility (Borowsky and Ellis 1987).

The bubble concept refers to a collection of smoke stacks or other pollution sources under single ownership. Regulation is concerned with total emissions exiting the bubble, i.e., the collection of sources owned by a single firm. This procedure will induce the firm to allocate emissions reductions across individual internal sources so as to obtain the least-cost reduction to the firm.

The ERC provision allows firms that reduce emissions by more than what is required by law to obtain a credit that can be "banked" for later use in bubble or offset or "netting" (see Borowsky and Ellis 1987). An ERC must be a surplus reduction that is "enforceable, quantifiable, and permanent" (Borowsky and Ellis 1987). Interstate trading of ERCs is permitted (Borowsky and Ellis 1987), provided that trades satisfy the conditions of the more stringent state. The ERC promotes efficient pollution control by inducing firms to reduce pollution more efficiently in order to earn the ERCs.

All of the approaches to implementing emissions reductions discussed so far have been based upon market incentives, in which firms have the incentive and flexibility to make choices that exhibit least-cost reductions. This makes the market-based policies attractive to economists; however, they have not always (hardly ever?) found favor with legislators and other groups in society. Reasons for this

8: Control Options and Strategies 111

opposition are discussed along with the political-economic aspects of designing and selecting control strategies (Chap. 9).

Comparisons of Market versus Command-and-Control Policies

The economists' preference for market-oriented policies was argued above on analytical grounds and based upon the presumption that greater flexibility generates more efficient (less costly) outcomes. This section considers the research that has tested the validity of this presumption.

Atkinson (1983) compared the effectiveness of market-oriented policies versus a command-and-control policy for controlling acid deposition in a simulation of the Cleveland region. He considered two versions of a marketable pollution right (MPR) scheme. One version is termed an emission discharge permit (EDP) system (which is equivalent to the EPS discussed in the previous section), while the other version is an ambient discharge permit (ADP) system. The command-and-control policy is the state implementation plan (SIP) found in the Clean Air Act. The SIP allocates emission reductions across all sources on an equal percentage-reduction basis. Atkinson found the ADP system to be more cost effective than the other systems in achieving local ambient standards. The annual costs for the three systems were (approximately) ADP, $83 million; SIP, $123.5 million; EDP, $158 million. The surprising result here is that the EDP system was more costly than the SIP command-and-control system, counter to the economist's expectations as argued above. The cost savings of the ADP and SIP systems come at the expense of increased long-range transport of acid deposition. This occurs because both the ADP and SIP schemes focus on local ground-level concentrations. Firms respond by building taller stacks to lower local concentrations and hence require fewer ADP permits and allow more complete oxidation of SO_2 downstream. Once the solution requires constraints on long-range transport, the cost advantage of the ADP is significantly reduced and the total emissions of SO_2 increase! By comparison, the total sulfate degradation for each of the systems was $11.25 \mu g/m^3$ for ADP; $8.36 \mu g/m^3$ for SIP; and $2.91 \mu g/m^3$ for EDP.

Atkinson's (1983) formal analysis also suggests that the cost-minimizing, marketable-permit approach necessarily leads to higher local pollution levels. McGartland and Oates (1985b) dispute the correctness of the formal analytical result. Using a counter example they show that the least-cost permit system does not necessarily lead to higher local pollution levels. However, at the practical level, they suggest that the Atkinson result may be valid. Their simulations of particulate-matter control in the Baltimore air quality–control region supported the Atkinson proposition – the cost-minimizing permit system generated higher local pollution levels than an EDP system. In response to McGartland and Oates, Atkinson (1985) concedes that his original proposition is stated too generally and offers a more restricted version in which the ADP costs are less than or equal to the SIP and EDP systems, since the ambient degradation of ADP will equal or exceed that of SIP/EDP systems.

A market-based policy proved more cost effective than a uniform percentage-reduction policy in a simulation of SO_2 reduction in Illinois (Diemer and Eheart 1988). The cost saving from using a transferable discharge permit (TDP) system was 40–60% of the cost of using a uniform reduction strategy. Their TDP system is equivalent to Atkinson's EDP system in that it is dependent upon emissions. Diemer and Eheart (1988) believe that their analysis is likely to underestimate the cost savings of TDP because of the restrictions they place on the market for permits. NSPS units are allowed to enter the market only as sellers of permits. The TDP system did result in some compliance violation but not as severe as the uniform reduction requirement. In general, the results depend upon the availability (and cost) of the nuclear power option as a means of reducing emissions.

Maloney and Yandle (1984) also found considerable cost savings from using an analog of the marketable permit system rather than less flexible control approaches, in a study of pollution control across different plants of the DuPont company based upon in-house data collection. They consider three regulatory approaches to achieving an overall 85% pollution abatement of hydrocarbon emissions. The first, termed source standards, requires 85% abatement for each source. The second approach, plant standards, allows the abatement to vary across sources within a given plant to achieve an overall plant abatement of 85%. This is similar to the EPA bubble. The final strategy considered allows the abatement to vary across plants as well to achieve an overall total plant abatement of 85%. This is termed the regionally marketable permit scheme.

Maloney and Yandle (1984) show that in moving from source

standards to plant standards a cost saving of 35% is achieved. The regionally marketable permit scheme achieves a cost saving of 76% over the rigid source standard approach. Using the marketable permit allows a 97% abatement for the same dollar cost as an 85% abatement using the source standards.

Morrison and Rubin (1985) used a linear-programming model to examine the effects of alternative emission control strategies for coal-fired power plants in the 31-state region of the eastern United States. Their framework allows for coal transportation costs as well as mine-mouth coal prices. Scrubbing and emissions-trading are considered control options for achieving 8-, 10-, and 12-million-tons-per-year reductions (below 1980 levels) in SO_2 emissions.

The analysis was designed to minimize the cost of delivered coal plus pollution-control costs subject to the constraints on energy demand and SO_2 emissions. In comparing the cost of the alternative control systems the authors concluded that the interstate emissions-trading strategy would reduce overall abatement costs by under 2%. A reduction in scrubbing costs of 14% was offset by 16% increase in the added cost of coal. The administrative complexity of the emissions-trading policy lead Morrison and Rubin (1985) to conclude that the additional cost saving was insufficient to warrant using such a trading policy.

Welsch (1988) considered the use of uniform reduction versus transferable emission permits in reducing SO_2 emissions in the British electric-power industry. He found that the transferable permits achieved an 88.9% reduction at a cost 27.1% lower than the uniform reduction approach. However, the nature of the FGD cost function dictates that the individual firms, given the option, either use FGD for all abatement, or use no active abatement at all, preferring instead to purchase permits.

References Cited

Atkinson, S.E. 1983. Marketable pollution permits and acid rain externalities. Can. J. Econ. 16:704–22.

———. 1985. Marketable pollution permits and acid rain externalities: A reply. Can. J. Econ. 18:676–79.

Borowsky, A.R., and H.M. Ellis. 1987. Summary of the final federal emissions trading policy statement. J. Air Pollut. Control Assoc. 37:798–800.

Carr, R.C. 1986. Integrated environmental control in the electric utility industry. J. Air Pollut. Control Assoc. 36:652–55.
Coal Week. 1987. September 14.
_____. 1988, November 7.
Dales, J. 1968. Pollution, Property and Prices. Toronto: University of Toronto Press.
Diemer, J.S., and J.W. Eheart. 1988. Transferable discharge permits for control of SO_2 emissions from Illinois power plants. J. Air Pollut. Control Assoc. 38:997–1005.
Gallogly, M.R. 1981–1982. Acid precipitation: Can the Clean Air Act handle it? Boston Coll. Environ. Aff. Law Rev. 9(3):687–744.
Hendricks, R.V., and P.S. Nolan. 1986. EPA's LIMB development and demonstration program. J. Air Pollut. Control Assoc. 36:432–38.
Jones, D.N., and R.A. Tybout. 1986. Environmental regulation and electric utility regulation: Compatibility and conflict. Boston Coll. Environ. Aff. Law Rev. 14:31–59.
Krupnick, A.J., W.E. Oates, and E. Van DeVerg. 1983. On marketable air-pollution permits: The case for a system of pollution offsets. J. Environ. Econ. and Manage. 10:233–47.
McGartland, A. 1988. A comparison of the marketable discharge permit systems. J. Environ. Econ. Manage. 15:35–44.
McGartland, A., and W.E. Oates. 1985a. Marketable permits for the prevention of environmental deterioration. J. Environ. Econ. Manage. 12:207–28.
_____. 1985b. Marketable pollution permits and acid rain externalities: A comment and some further evidence. Can. J. Econ. 18:668–75.
Maloney, M.T., and B. Yandle. 1984. Estimation of the cost of air pollution control regulation. J. Environ. Econ. Manage. 11:244–63.
Maulbetsch, J.S., M.W. McElroy, and D. Eskinazi. 1986. Retrofit NO_x control options for coal-fired electric utility power plants. J. Air Pollut. Control Assoc. 36:1294–98.
Montgomery, W.D. 1972. Markets in licenses and efficient pollution control programs. J. Econ. Theory 5:395–418.
Moore, T. 1987. How advanced options stack up. EPRI J. (July/Aug.):4–13.
Morrison, M.B., and E.S. Rubin. 1985. A linear programming model for acid rain analysis. J. Air Pollut. Control Assoc. 35:1137–48.
Muzio, L.J., and G.R. Offen. 1987. Assessment of dry sorbent emissions technologies. Part I. J. Air Pollut. Control Assoc. 37:642–54.
Offen, G.R., D. Eskinazi, M.W. McElroy and J.S. Maulbetsch. 1987. 1987 Joint EPRI/EPA symposium on stationary combustion NO_x control. J. Air Pollut. Control Assoc. 37:864–71.
Purdy, P. 1988. "Baking soda helps clean up acid rain." *Denver Post,* 9 October 1988, sec. G.
Regens, J.L., and R.W. Rycroft. 1986. Options for financing acid rain controls. Nat. Res. J. 26:519–49.

Shepard, M. 1988. Coal technologies for a new age. EPRI J. (January/February): 4–17.
Streets, D.G., and T.D. Veselka. 1987. Economic incentives for the reduction of sulfur dioxide emissions. Energy Syst. Policy 11:39–59.
Welsch, H. 1988. A cost comparison of alternative policies for sulphur dioxide control: The case of the British power plant sector. Energy Econ. (October):287–97.

CHAPTER 9

Exaggeration of Control Costs and Disruption

Introduction

Chapter 8 discussed alternative control strategies for reducing precursor emissions of acid rain. This chapter deals with estimates of the costs of abatement. The opponents of acid rain controls emphasize the uncertainty of benefits from control, while claiming firmness of the high costs of control and economic disruption that would result from such controls. Forster (1989) argued that the controls opponents have incentives to exaggerate the costs of control and economic disruption to avoid or delay a controls program. Forster (1989) not only demonstrated a wide range of cost estimates, but showed that there were conflicting analyses of impacts consistent with the incentive hypothesis. The following discussion follows and expands upon Forster (1989).

As has been noted frequently, the major target of acid rain control proposals in the United States is the sulfur dioxide (SO_2) emissions from the electric-power plants that are exempted from the Clean Air Act (CAA) of 1970 and the 1977 amendments to the CAA. As noted in the previous chapter, the 1977 amendments essentially require new power plants to use scrubbers to abate SO_2, either through a uniform percentage reduction requirement (New Source

9: Exaggeration of Control Costs and Disruption

Performance Standards, NSPS, amendment) or through the best available control technology (BACT) requirement (prevention of significant deterioration, PSD, amendment). Any power plant already operating or in the construction phase when these regulations came into effect was exempted through a "grandfather" clause. The majority of these power plants are concentrated in the states of the upper Ohio Valley, which is a major regional source of SO_2 emissions as noted in Chapter 2.

Any acid rain control program is likely to increase the production costs for this group of plants. This will happen even if the intervention permits the freedom-of-choice abatement option rather than mandating a BACT approach (i.e., scrubbers). The electric-power companies will respond to these cost increases by shifting them to other economic groups. The most obvious response is to shift the cost increase forward to electricity consumers in the form of higher electricity prices. Less obvious perhaps is a shift backward to the factors of production. Most important in this group are the coal companies that supply the utilities with coal. The use or price of their coal may have to fall, for example, since with other production costs rising, the electric utilities may not be able to continue using as much coal at the prevailing price prior to the control requirement. Any drop in coal price or usage will affect the coal companies' factors of production, mainly the coal miners.

Four groups of economic agents are usually on opposite sides of economic bargaining: utility companies, electricity consumers, coal companies, coal miners. Electricity consumers want low electricity prices, while the utilities want high prices for electricity but low coal prices. The coal companies prefer high coal prices and lower wages, while the coal miners want higher wages. However, the threat of acid rain controls for the electric utilities forges an informal coalition between the four agents. The members of this informal coalition have a common interest in avoiding or delaying any acid rain program to avoid wealth losses.

The coalition has an incentive to exaggerate impacts of controls and to emphasize uncertainty of benefits given Executive Order 12,291, which was discussed in Chapter 1. While in theory it is possible to design mechanisms that remove this incentive (Kwerel 1977; Baron 1985), the stance of the Reagan administration and the burden-of-proof requirements embodied in Executive Order 12,291 suggested that the exaggerated costs would be accepted as credible.

Electric Utilities

Those electric utilities that were "grandfathered" out of the 1977 CAA amendments have been receiving special economic benefits. Those amendments serve as a barrier to entry into the electric-power-generation industry. The cost structure facing new entrants reflects their higher cost of required emissions-control technology. An acid rain controls program would reduce the competitive advantage and the profitability of those existing grandfathered plants. As an example of the benefits from such barriers to entry, the PSD requirement may have increased the return on non-ferrous smelting in portfolios by 44% for existing smelting operations (Maloney and McCormick 1982).

As has been noted previously the amendments requiring "uniform percentage emissions reductions," or BACT, eliminate any incentive for coal-fired plants to switch to low-sulfur coal. Gallogly (1981–1982) suggests that the intent of these amendments, among other things, was to free low-sulfur coal for the older plants to satisfy the 1970 CAA requirements. Crandall (1983), in a more cynical vein, concluded that Congress deliberately laid a disproportionate burden for air quality control on the southern and western regions of the country. The PSD policy, for example, protects the older plants in the so-called "Frost Belt" (i.e., the northern and midwestern regions) by making it more costly for firms to move westward or southward to the "Sun Belt."

Despite the impression that control costs are better known than control benefits, the published, annual control-cost estimates for the electric-power industry exhibit a wide range of values. There are several factors that will produce legitimate variation in cost estimates, including the size of the required emissions reductions and the method specified for achieving the reduction. However, the conditions underlying the stated cost estimates may be unspecified or, if specified, selected to produce exaggerated costs.

A senior vice-president of the American Electric Power Service Corp., Mr. Joseph Dowd (1982), claims that a 5% reduction in emissions would cost the utilities $3–4 billion annually for capital costs, while a 30% reduction hikes the estimate to $12–14 billion. An additional $0.5–4 billion must be included for operating costs. These figures are in marked contrast to those of the Office of Technology Assessment (OTA 1984). According to the OTA assessment a 4.6-

million-ton-reduction in SO_2 would cost $0.6–0.9 billion using the most cost-effective method, while scrubber technology would cost $1.4 billion for the same reduction (all in $ 1982). Since 25-million tons of SO_2 were emitted in 1980, a 4.6-million-ton, or roughly a 20%, reduction in SO_2 according to the OTA, is cheaper than the estimate of 5% by Dowd. For an 11.4-million-ton reduction, OTA estimates a least cost of $4.2–5.0 billion and a scrubber cost of $5.9 billion. An 11.4-million-ton reduction would cost less than half of Dowd's estimate for a 30% reduction, or roughly 8-million tons.

Using a computable, general equilibrium analysis that allows for inter-industry output and price adjustments, Willett (1986) determined that a 5% SO_2 emissions reduction would cost $0.191 billion, while a 25% reduction would cost $0.954 billion. If all electric utilities were required to use BACT, the annual costs would be $4.053 billion. Willett's cost figures are in 1972 dollars, and adjusting for inflation, they are still lower than Dowd's.

The Edison Electric Institute (EEI) estimated the cost of the 1987 Gregg-Downey sulfur and nitrogen emissions (SANE) tax bill (H.R. 2497) to be $16.3 billion (*Coal Outlook* 1987). This estimate, which was also endorsed by the National Coal Association, is in the upper range of estimates scattered throughout the literature and strikes an economist as surprising, since this, as argued in the previous chapter, should produce a "least-cost" approach. However, EEI, in making this calculation, assumed that emissions would remain at their current levels and the tax paid on these emissions. This assumes that no electric utilities could make any adjustments to lessen the tax burden. This of course would mean that the tax has failed in its intent! The utilities are unlikely to continue emissions at current levels in the face of such a tax. However, by ignoring such adjustments available to producers, the control cost is exaggerated and the tax becomes a less attractive policy option. The EPA estimated that the tax would lead to a 75-million-ton-per-year switch from high-sulfur to low-sulfur coal.

Coal Outlook (1987c) offered annual control-cost estimates ranging from $3 billion to $25 billion with no specifications concerning size or method of reduction. This is a very wide band. The Industrial Gas Cleaning Institute (IGCI) claims that the cost of installing scrubbers has been greatly overestimated. In their assessment, a 10- to 12-million-ton reduction in SO_2 would have a first-year cost of $3.4 billion and could be as low as $1 billion (*Coal Outlook* 1987b).

Since the IGCI represents the manufacturers of scrubbers, it has an incentive to under-represent the cost of control to improve the chances that a BACT program will be successful.

The previous discussion shows that rather than having known-cost estimates, there is a wide range of estimates. This wide range may be explained by the incentives that economic agents have to exaggerate cost information to their best advantage.

Occasionally industry representatives argue that special acid rain controls are not necessary, since the CAA and time will solve the problem. As old plants are retired they will be replaced by new plants, which will be subject to the NSPS amendments. However, this ignores the fact that the NSPS induces firms to extend the life of the old plant beyond its otherwise normal retirement date. Utilities were not ordering new coal plants in the 1980s. Instead the utilities were "using and rejuvenating conventional coal-fired plants they have, and increasingly, restarting the oil and gas-fired plants they shut down following the crises of the 1970's" (*Coal Age* 1987). The cost of extending the life of an existing facility is between $125/kW and $625/kW compared to a range of $1300/kW to $1900/kW for a new plant (Douglas 1987). As long as the cost of rejuvenation is less than 50% of the cost of a new plant, then the firm escapes the NSPS. It is not clear whether the 50% rule is cumulative, or applies to one-time rejuvenation (Douglas 1987).

In late 1988, the EPA determined that the plans of Wisconsin Electric Power Company (WEPCO) to rejuvenate its already 50-year-old Port Washington power plant would be subject to the NSPS. WEPCO's proposed renovations included replacing rear steam drums on four out of five units, repair or replacement of air heaters, and renovation of major mechanical and electrical systems (*Coal Week* 1988a). WEPCO claimed these proposals were "maintenance" and hence did not come under the NSPS. EPA disagreed declaring the changes were "major modifications," which would increase emissions of SO_2 and NO_x (*Coal Week* 1988a). WEPCO countered by filing suit with the Seventh U.S. Circuit Court of Appeals in Chicago (*Coal Outlook* 1988b).

This case was very important and other utilities watched to see how EPA would handle the cases. EPA claimed that they would handle the issue on a case-by-case basis (*Coal Outlook* 1988c). The coal and electric-power industries expressed concern that EPA's determination of the WEPCO proposal may lead to the shutdown of some of the smaller, older units, while others may seek alternative energy sources such as gas or Canadian power (*Coal Week* 1988a). In the end, a compromise was worked out between EPA and WEPCO (*Coal Week* 1992). This compromise, known as the WEPCO provision or the WEPCO fix, will also be used by EPA in applying the standards of the 1990 Clean Air Act (see Epilogue). Modifications that do not result in an increase in emissions will not trigger the new source standards.

If acid rain control is going to be required, there are several institutional and economic factors that would lead the utilities to select scrubbers rather than alternatives, such as switching to low-sulfur coal, for example. Electric utilities are subject to rate-of-return regulation by public utility commissions (PUCs). This means that the utilities' allowable profits are restricted to be no greater than a specified percentage of their invested capital. Pollution-control devices such as scrubbers expand the utilities' capital base and hence allow them to earn higher profits by obtaining price increases. In many cases the utilities may even receive credit for construction work in progress (CWIP) during the installation of pollution-control measures, although 12 states prohibited CWIP in 1985 (Jones and Tybout 1986). A report from the National Coal Council to the Energy Department advocates allowing full credit for CWIP (*Coal Outlook* 1988a). In a unique program the state of Ohio is permitting firms to recover costs for coal research and development every 6 months through requests for gas or electricity rate hikes (*Coal Age* 1987). This feature of the regulatory process creates a bias in favor of capital-intensive pollution control, which will increase the social cost of controlling emissions (Tschirhart 1984).

A further bias toward scrubber technology may be created through tax policy. According to Jones and Tybout (1986), the major group benefiting from pollution-control tax breaks is the electric-power industry. They point out that since, in essence, the taxpayer is paying the bill through tax breaks given to the industry there is no incentive for the industry to undertake efficient pollution control or to develop or adopt new cost-saving technologies.

The tax concessions available to the utilities for pollution control may be in the form of investment tax credits, accelerated cost recovery, or industrial development bonds. These tax concessions were only granted for identifiable pieces of equipment (Jones and Tybout 1986), which created a bias in favor of "end-of-pipe" technologies such as scrubbers rather than "in-process" technologies. Furthermore, PUC accounting practices ensured that the benefits of these tax concessions accrue to the utilities rather than the rate-paying public (Jones and Tybout 1986). As a result of these tax concessions for pollution control, the effective tax rate for the electric utilities was one-fifth the average for the corporate sector between 1964 and the late 1970s, while the rates were roughly equal in the mid-1950s (Jones and Tybout 1986). In dollar terms, the benefits to the industry from these concessions amounted to annual savings in excess of $4.3 billion by the late 1970s (Jones and Tybout 1986). Tax reform in 1986 removed these tax concessions for pollution control.

In what initially appears paradoxical, in the summer of 1983, a group of electric utilities calling itself the American Public Power Association broke ranks with the rest of the electric utilities and advocated a federal acid rain controls program. Their proposal was based upon an emissions tax on all U.S. emitters, not just the electric utility industry (*Coal Week* 1983b). The association comprised 1750 power plants, most of which were small and publicly owned. Electric Power Research Institute (EPRI) research shows that small power plants would find flue-gas desulfurization (FGD) costs very high and would prefer to switch fuels to reduce emissions (Peck 1986) (see Table 9.1). Thus the move by the American Public Power Association is not so surprising after all.

Another special-interest group calling for a flexible approach to acid rain control is the American Gas Association (AGA), which advocates permitting utilities to choose the most cost- and environmental-effective method of emissions reduction. Their policy calls for a 10-million-ton reduction in SO_2 and a 4-million-ton reduction in NO_x (*Coal Week* 1988b). Of course, a flexible approach would permit a switch in fuel to natural gas. However, the AGA expects that most companies would "co-fire" rather than completely switch fuels (*Coal Week* 1988b).

In another study the AGA analyzed 15 control technologies and found that 6 could achieve significant reductions in SO_2 and NO_x; 5 of the 6 use gas. The most cost effective was combined-cycle repowering, which achieved a reduction of 100% in SO_2, 90% in NO_x, and 60% in CO_2. The most expensive was the scrubber technology (*Coal Outlook* 1989).

Coal Suppliers

The coal that is burned by the utilities in the upper Ohio Valley comes largely from local sources and has a high sulfur content. The coal companies in this region are earning special profits, which economists call "rents," compared to the western low-sulfur coal companies. There are three separate pieces of legislation that have improved the position of eastern and midwestern coal producers relative to their western competitors (*Coal Week,* 1984a).

First, the Clean Air Act Amendments of 1977 neutralized the advantages of using low-sulfur coal to reduce emissions of SO_2 that prevailed prior to 1977 by imposing a uniform percentage reduction requirement for new sources. Not only does the NSPS essentially require a technology solution, but Reinert and Ratick (1988) show that the long-run average costs of FGD processes decline with the coal's sulfur content and the level of SO_2 removed. The second piece of legislation is the Staggers Act, which permits the railways to essentially charge monopoly rates that cause the price of western "delivered" coal to rise. According to the National Coal Association (NCA) 85% of coal carried by rail is "captive" of a single rail transporter (*Coal Week* 1984b). If rail rates were more competitive, then the price of delivered western coal could drop and more could be sold. It has been estimated that 15% more Wyoming coal could be sold if rail rates were competitive. Unit-train transportation charges can account for up to 75% of the delivered price of Wyoming coal, and in the early 1980s coal rail rates in the Powder River Basin exceeded marginal cost by 112% (Atkinson and Kerkvliet 1986, correction by author). In recent years the competition to Burlington Northern by the entry of the Chicago and North Western Railway into the Powder River Basin has caused substantial decreases in rail rates, including some as much as $7 per ton (BLM 1987).

The third piece of legislation is the Surface Mine Control and Reclamation Act of 1977. This act increases the cost of strip-mining coal and gives an advantage to underground mining in the Midwest and East. Kalt (1983) found that the redistribution of income following this third piece of legislation resulted in increases of $130 million (per year) for underground-coal producers and $1.1 billion for the consumers of environmental amenities. Kalt estimated losses of $400 million for coal consumers and losses of $980 million for surface-coal producers. In view of these redistributions of wealth, it is not surprising that a coalition of 19 low-sulfur coal companies formed the Alliance for Clean Energy (ACE) and split away from the NCA during the summer of 1983 to advocate acid rain control legislation that permitted freedom of choice in the method of emissions reductions (*Coal Week* 1983a). The ACE argues this would be a more cost-effective solution to acid rain controls—which they see as inevitable—than a BACT approach (*Coal Week* 1983a). The stand by ACE not only angered the NCA, but one of its major customers, the American Electric Power Co., issued letters virtually threatening their buyer-seller relationship (*Coal Outlook* 1984).

Some capital-intensive coal producers take a long-term view of the coal market, view acid rain controls as inevitable, and are investing in low-sulfur reserves. Labor-intensive companies are not taking a long-term approach (*Coal Outlook* 1987a).

The disruption to high-sulfur coal markets caused by fuel-switching as an alternative under freedom of choice may not be as great as suggested by the coal producers, although they do have an incentive to exaggerate the magnitude of the disruption. The extent of shifting has been estimated to be between 50- and 130-million tons annually (Peck 1986). As fuel switching takes place it puts upward pressure on the price of low-sulfur coal, thus reducing the relative attractiveness of this option compared to using scrubbers. This permits continued use of local high-sulfur coal. Failure to account for such price changes will exaggerate the extent of fuel switching and consequent disruption. "Very low" sulfur coal could rise in price by $11 per ton and low-sulfur coal could rise in price by $1–8 per ton (*Coal Outlook* 1987h).

Recent results suggest that the big winner from fuel switching as an acid rain control would be the low-sulfur coal producers of central Appalachia (*Coal Outlook* 1987e) with perhaps negligible impacts on Western coal fields (*Coal Outlook* 1987h). The losers would be northern Appalachian and midwestern high-sulfur coal producers (*Coal Outlook* 1987e).

Table 9.1 shows the factors influencing the decision to switch or scrub.

EPRI research shows that the extent of fuel switching versus scrubbing depends upon the size of the mandated reduction in SO_2. According to one report, scrubbing accounts for 30–47% of a 5-million-ton reduction, 50–66% of an 8-million-ton reduction, and

Table 9.1. Factors in utility scrub-switch decisions

Factor	Impact	Decision Favored
Near high-sulfur coal production	Low transportation costs for high-sulfur coal	Scrub
Coal delivered by barge	Low transportation costs for low-sulfur coal	Switch
Cyclone boiler	Limited low-sulfur coal options	Scrub
Small units	High FGD costs	Switch
New units	Long FGD lifetime	Scrub
Congested site	High FGD costs	Switch
Baseload unit	Low FGD costs	Scrub
Cycling unit	High FGD costs	Switch
Unit burns mid-sulfur coal	Small SO_2 reduction by switching	Scrub
High real discount rates	Low levelizing factor	Switch
Tax-exempt financing	Low FGD financing costs	Scrub
High corporate debt	Limited financing capability	Switch
Utility owns high-sulfur coal mines	Low incremental cost for high-sulfur coal	Scrub
Increased rail competition	Lower transportation	Switch
High low-sulfur coal prices	High switching costs	Scrub

Source: Peck (1986).
Note: FGD = flue-gas desulfurization.

65–75% of a 10-million-ton reduction (Platt 1986). In this study Western coal received a negligible boost. A more recent report on this EPRI study suggests that for small SO_2 reductions the utilities would still choose to scrub rather than disrupt their fuel supply patterns (*Coal Week* 1987b). Requirements for large SO_2 reductions would push up the price of very low sulfur coals, and moderately low sulfur coals would not be sufficient to achieve the desired reductions. According to EPRI forecasts, a $2 per ton increase in the price of "compliance" coal would reduce its consumption between 15- and 60-million tons per year, with a suggested average of 30-million tons.

A further bias against the Western region was revealed in the fall of 1988 when the Federal Energy Department approved 16 clean-coal projects, none of which were from the West (except one from Texas). Two Wyoming projects were rejected. The accepted proposals also favored "retrofit" technology for existing plants rather than proposals for new fuels (*Casper Star-Tribune* 1988a). This approach favors the midwestern and eastern coal producers.

Coal Miners

Losses to the coal producers will be passed backward to the mine workers in the adversely affected regions. The United Mine Workers of America (UMWA), according to its president, R.L. Trumka (1984), "sincerely believe that the issue of acid rain must be met head-on but it must be done on the basis of solid scientific evidence." The UMWA continues to argue that the science on acid rain effects is uncertain, with conflicting analyses (United Mine Workers Journal 1988). The earlier chapters on impacts demonstrated that there is a basis for such a claim. However, when the UMWA asserts that the National Acidic Precipitation Assessment Program (NAPAP) interim reports state that "there is absolutely no hard scientific evidence" of "aquatic damage or structural damage due to long-range transport of sulfur dioxide," it is distorting or misreporting the results.

The UMWA is concerned with the loss of jobs for coal miners if legislation is introduced permitting fuel switching (Meyer and Yandle 1987). In order to save jobs the UMWA argues that if legislation is forthcoming, it should be a technology-based strategy such as is embedded in the uniform percentage reduction concept. "The concept of 'percentage reduction' will save thousands of jobs and will allow

9: Exaggeration of Control Costs and Disruption

our nation's industrial base to continue to use relatively inexpensive local coal while moving toward cleaner air . . ." (Trumka 1984). Yandle (1985) argues that a coalition of eastern coal miners, environmentalists, and regional political interests were responsible for obtaining the 1977 amendments to the CAA, which reduced the advantage of burning low-sulfur western coal. Yandle further points out that the western coal miners were not "organized workers," as were the eastern miners who were successful in obtaining legislation that provided protection from competition for their employers as well as themselves (Yandle 1985).

For the acid rain issue in the 1980s, the UMWA calculates that if only 50% of the reductions in emissions is achieved through fuel switching (to low-sulfur coal), at least 39,000 high-sulfur-coal-mining jobs would be lost. With 75% of the reductions being achieved through fuel switching, the job losses rise to 60,000. They point out that this will lead to other non-coal job losses as well (Trumka 1984). The UMWA tends to refer to the job losses in the adversely impacted regions and ignores the expansion in jobs that will occur in the low-sulfur coal mines of central Appalachia and the West. The OTA (1984) study suggests that "freedom-of-choice" legislation would reduce jobs in the high-sulfur regions by 20,000–30,000, with an equivalent expansion of jobs in the low-sulfur regions. Willett (1986) determined overall expansions in output in the mining sector.

The recent UMWA estimate of job loss to the coal-mining sector from the Waxman-Sikorski (H.R. 2666) and Mitchell (S. 1894) acid rain bills is 40,000 mining jobs in the Midwest and northern Appalachia (United Mine Workers Journal 1988). Once again they ignore the coal-mining, job-creation potential in the low-sulfur-coal fields. One coal executive is quoted as saying that acid rain controls would reduce jobs by 1500 in northern West Virginia but expand jobs by 16,000 in southern West Virginia (*Coal Week* 1988b)!

The UMWA calculations exaggerate the disruption to the coal miners by ignoring the job expansion in low-sulfur coal regions of the country. Consideration of such disruption is indicated in Executive Order 12,291. The UMWA is more concerned with protecting the jobs of the coal miners east of the Mississippi where their membership is heaviest. A new job created in the western region of the country is not a perfect substitute for the job lost in the eastern region in the mind of the displaced miner, nor are local opportunities attractive to the UMWA. As one official put it, "We can't give up someone's job in the hope he can go somewhere else and get a job, whether it be in southern West Virginia, eastern Kentucky or anywhere else" (*Coal Week* 1988b). There are adjustment costs that face the displaced miners in changing geographic locations as well as changing occupations.

Not surprisingly the UMWA supports, as the "responsible acid rain bill," Senator Byrd's bill, S. 879, which would devote $3.5 billion over a 10-year period to the development of clean coal technology (United Mine Workers Journal 1987, 1988).

Electricity Consumers

The impacts of acid rain controls will not only be shifted backward onto the coal producers and coal miners, they will be shifted forward to electricity consumers. To the extent that the cost of control to the electric utilities is overstated, so will the increase in electricity rates be overstated. This will increase the opposition of electricity consumers. The National Coal Association estimated that the SANE tax would increase industrial consumer electricity rates by 13% (*Coal Outlook* 1987f). This estimate is based upon the EEI calculations that assumed no change in emission levels, and hence is biased upward.

A major electricity-consuming coalition of the National Association of Manufacturers (NAM) is opposed to acid rain legislation (*Coal Outlook* 1987c). A study done for NAM by Data Resources Inc. (DRI) estimates that stringent acid rain legislation would boost consumer electricity prices by 13%. An industry lobby group called the Citizens for Sensible Control of Acid Rain raised $4.8 million to fight acid rain controls. They claim that electricity prices will jump 30% (*Environmental Action* 1987). However, Diemer and Eheart (1988), in their simulation of acid rain controls in Illinois, found that electricity rates would rise by less than 2%!

The study for NAM conducted by DRI also estimated that stringent acid rain legislation would result in (1) a loss of 862,000 jobs in the United States, (2) a drop in gross national product of $223 billion, and (3) a worsening of the balance of payment (*Coal Outlook* 1987i). These results are in marked contrast to a study in early 1987 by Management Information Services, which concluded (1) jobs would increase by 100-194,000; (2) the national economy would improve by $7.5-13 billion; and (3) the trade deficit would improve (Goeller

1987). DRI concedes that it ignored some of the positive benefits that would occur (*Coal Outlook* 1987i).

A study by the IGC concluded that acid rain legislation would create 600,000–person years of employment (Smith 1988). The study did not consider the loss of jobs that might result; hence, it makes acid rain controls appear more favorable in a benefit-cost calculation. As mentioned earlier, the IGCI has a natural incentive to downplay the cost side of the acid rain control issue.

Regional Political Representatives

The intricacies involved in obtaining legislative approval for acid rain controls in Congress are detailed by McPoland (1985). According to former EPA Administrator William Ruckelshaus his main frustration at EPA in 1984 was "the politicization of the acid-rain issue. On top of an inherently emotional problem, we have to deal with politics with a big P. . . . Political pressures are increasing and that tends to erode the trust of the American public" (U.S. News & World Report 1984).

The behavior of the politicians on the acid rain issue is consistent with the capture theory of economic regulation in which the politicians respond to the special economic interest groups of the regions that they represent. The opponents of acid rain controls, however, couch their arguments in terms of scientific uncertainty, cost, and disruption to local economies. For example, Congressman Rahall (D. W. Va.) argues, "From my understanding of the acid rain issue, I do not believe that there is sufficient scientific evidence to support a massive and costly control program . . ." (Rahall 1984). Others from the upper Ohio Valley who have been outspoken in their opposition to a controls program include Congressmen Luken (D. Ohio) and Dingell (D. Mich.), and Senators Byrd (D. W.Va.) and McConnell (R. Ky.). Congressman Eckart (D. Ohio) had the crucial vote that killed a controls legislation package in 1984. Until this vote, Eckart was considered sympathetic to environmental issues and likely to support the legislation. As chairman of the House Committee on Energy and Commerce, Dingell was a particularly powerful opponent. In early 1987, Dingell stacked his committee by adding three new members who shared his views on acid rain (Engelau 1987). Rahall and Byrd have warned that there may be a hidden agenda to undermine coal as an energy resource. Rahall (1984) has also suggested that there is a threat to the U.S. industry from Canadian utilities:

> I also fear that an acid rain control program of any kind may serve as a disincentive to the construction of new coal-fired utility units and the conversion of oil-fired boilers to coal. Witness the increased use of Canadian power by utilities in the Northeast since the acid rain bandwagon began to roll. The Canadians have constructed facilities with excess power and are aggressively selling it in the U.S. and Canadian power imports—largely nuclear and hydro-based—will soon displace the equivalent of 25 million annual tons of coal consumption in the U.S.

This same view has been stated by the governor of West Virginia, Arch Moore (R). Moore (1987) is quoted as telling the American Mining Congress' 1987 Coal Convention:

> The objective of the Canadians—we know . . . their objective is to reduce the amount of U.S. coal that is burned period. . . . Once they've succeeded in putting high-sulfur coal out of business they are going to put their efforts on low-sulfur coal.

The Canadian electricity threat shows an interesting special interest schism in the unholy alliance. In order to counter electricity imports from Canada, the U.S. electric industry proposed that an electricity import tax be imposed. The tax was not supported by the NCA, which pointed out that the imports generated benefits for U.S. consumers by displacing higher-cost American production. The NCA also points out that they sell more than 3-million tons of coal annually to be used in Canadian electricity production (*Coal Outlook* 1987d). A tariff on Canadian electricity imports would jeopardize this market for the members of the NCA.

The politicians who have proposed acid rain controls legislation either tend to come from acid rain impacted areas, or have designed legislation that is to the advantage of their own region.

Those proponents of legislation from impacted areas include Congressman Sikorski (D. Minn.), Senator Durenberger (R. Minn.), Senator Stafford (R. Vt.), and Senator Mitchell (D. Maine). Senator Mitchell's bill finally failed in October 1988 when the senator withdrew the bill in the face of a filibuster threat. Senator Byrd and the UMWA had been attempting to work with Mitchell on a "compromise" bill. The UMWA wanted a 10-million-ton reduction using scrubber technology phased in over a 15-year period. The bill was defeated in part by opposition from western senators. Senator Simpson (R. Wyo.) stated: "I'm not going to accept any kind of acid rain bill that gives advantage to high-sulfur coal over low-sulfur coal . . ." (*Casper Star-Tribune* 1988b). Even the Clean Air Coalition, an environmental lobby, rejected the bill preferring no bill to a weak one (*Casper Star-Tribune* 1988b).

The special interests of the western states were evident in an earlier bill sponsored by Congressmen Cheney (R. Wyo.) and Udall (D. Ariz.). Their bill called for freedom of choice in meeting emissions reductions rather than mandating scrubber technology.

The special interest of the high-sulfur coal states is embodied in Senator Byrd's proposed legislation, which promotes the development of clean-coal technology. In addition to moving toward a "technology" solution this bill would delay "active" acid rain control with its emphasis on "developing technology." Senator Byrd's bill calls for an expenditure of $3.5 billion over a 10-year period.

The state governments of West Virginia and Ohio are sponsoring clean-coal development for their electric power industries. In West Virginia, tax structure changes, simplified permitting, and the provision of incentives are part of a plan to build fluidized-bed combustion plants. Ohio's Coal Office supplied seed money for clean-coal demonstration projects (*Coal Week* 1987a). In 1987 a coalition of Northeast and Midwest congressmen formed to advocate special steps to protect the interests of the high-sulfur-coal producer (*Coal Outlook* 1987g).

Senator Byrd was a particularly powerful opponent of acid rain control legislation during the 1980s, given his position as Senate majority leader. In this capacity he had the ability to control the Senate agenda. In late 1988 Senator Byrd stepped down from this position. Senator Mitchell was elected in his place as Senate majority leader. Given Senator Mitchell's past support for acid rain controls, this appeared to increase the probability for successful legislation. However, *Coal Outlook* (1988b) said that Mitchell should not be labelled as dogmatic on the acid rain issue. In his new role of Senate majority leader he may be even more aware of the importance of achieving a consensus approach.

References Cited

Atkinson, S.E., and J. Kerkvliet. 1986. Measuring the multilateral allocation of rents: Wyoming low-sulfur coal. Rand J. Econ. 176:416–30.

Baron, D.P. 1985. Regulation of prices and pollution under incomplete information. J. Public Econ. 28:211–31.

Bureau of Land Management (BLM). 1987. Case study: Impact of coal transportation on western coal development and the federal coal program. Washington, D.C.

Casper Star-Tribune. 1988a. September 29.

———. 1988b. October 5.

Coal Age. 1987. March 17, pp. 11–12.

Coal Outlook. 1984. May 28.

———. 1987a. August 10.

———. 1987b. August 17.

———. 1987c. August 24.

———. 1987d. August 31.

———. 1987e. September 7.

———. 1987f. September 14.

———. 1987g. September 21.

———. 1987h. September 28.

———. 1987i. October 12.

———. 1988a. November 28.

———. 1988b. December 5.

———. 1988c. December 12.

———. 1989. January 30.

Coal Week. 1983a. August 22.

———. 1983b. August 29.

———. 1984a. April 16.

———. 1984b. September 10.

———. 1987a. May 11.

———. 1987b. September 8.

———. 1988a. October 31.

———. 1988b. December 12.

———. 1992. May 25.

Crandall, R.W. 1983. Controlling Industrial Pollution. Washington, D.C.: Brookings Institution.

Diemer, J.S., and J.W. Eheart. 1988. Transferable discharge permits for control of SO_2 emissions from Illinois power plants. J. Air Pollut. Control Assoc. 38:997–1005.

Douglas, J. 1987. Longer life for fossil fuel plants. EPRI J. (July/August): 20–27.

Dowd, J.A. 1982. Comment during question period 2. In Acid Rain: A Transjurisdictional Problem in Search of a Solution. Ed. P.S. Gold, p. 118. SUNY at Buffalo: Canadian-American Center.

Engelau, D. 1987. Washington Report. J. Air Pollut. Control Assoc. 37:776.

Environmental Action. 1987. In sheep's clothing. September/October: 9.

Forster, B.A. 1989. Acid rain games: Incentives to exaggerate control costs and economic disruption. J. Environ. Manage. 28:349–60.

Gallogly, M. 1981–1982. Acid precipitation: Can the clean air act handle it? Boston Coll. Environ. Aff. Law Rev., 9(3): 687–744.

Goeller, D. 1987. Acid rain controls would boost U.S. economy, study says. Associated Press, *Laramie Boomerang,* February 1, p. 23.

Jones, D.N., and R.A. Tybout. 1986. Environmental regulation and electric utility regulation: Compatibility and conflict. Boston Coll. Environ. Aff. Law Rev. 14:31–59.

Kalt, D. 1983. The costs and benefits of federal regulations of coal strip mining. Nat. Resour. J. 23:893–915.

Kwerel, W. 1977. To tell the truth: Imperfect information and optimal pollution control. Rev. Econ. Stud. 44:595–601.

McPoland. F.F. 1985. Acid rain: Legislative perspective. In Acid Deposition: Environmental, Economic, and Policy Issues. Ed. D. A. Adams and W.P. Page, pp. 453–66. New York: Plenum Press.

Maloney, M.T., and R.E. McCormick. 1982. A positive theory of environmental quality regulation. J. Law Econ. 25:99–123.

Meyer, R., and B. Yandle. 1987. The political economy of acid rain. Cato J. 7:527–45.

Moore, A. 1987. Presentation before the 1987 Coal Convention of the American Mining Congress. Coal Week, May 11.

Office of Technology Assessment (OTA). 1984. Acid rain and transported air pollutants: Implications for public policy, U.S. Congress. OTA-0-204, Washington, D.C.

Peck, S.C. 1986. The role of low-sulfur coal in SO_2 reduction strategies. EPRI J. (December):53–55.

Rahall, N.J. 1984. The congressional debate. J. Air Pollut. Control Assoc. 34 (6):625–26.

Reinert, K.A., and S.J. Ratick. 1988. A note on estimating a long-run average cost curve for flue gas de-sulfurization. J. Environ. Econ. Manage. 15:35–44.

Smith, J.C. 1988. IGCI newsletter. J. Air Pollut. Control Assoc. 38:878.

Trumka, R.L. 1984. Luncheon address. J. Air Pollut. Control Assoc. 34 (6): 628–30.

Tschirhart, J.T. 1984. Transferable discharge permits and profit-maximizing behavior. In Economic Perspectives on Acid Deposition Control. Ed. T.D. Crocker, pp. 157–71. Stoneham, Mass.: Ann Arbor Science.

United Mine Work. J. 1987. September.

_____. 1988. April.

U.S. News & World Report. June 18, 1984. After the turmoil at EPA—the successes and the failures. Interview with W. Ruckelshaus.

Willett, K. 1986. Environmental management costs using a best available control technology BACT in the Electric Generating Industry. Managerial Decis. Econ. 7:29–36.

Yandle, B. 1985. Unions and environmental regulation. J. Labor Hist. 6:429–36.

CHAPTER **10**

Scientific Validity and Political Sensitivity

Introduction

Chapters 2 through 9 show that there is considerable scientific uncertainty concerning not only the benefits of acid rain control but also various other aspects, including the costs of controlling acid rain precursor pollutants.

The opponents of acid rain controls use the uncertainty of potential benefits of control as an argument to delay a controls program while calling for further research into the impacts of acid rain. However, there are two major reasons to believe that this call for further research to resolve these impact uncertainties may not be the socially optimal response. The first concerns problems in incentive structures that may affect the research thrust, and the second concerns salient qualitative features of the physical impacts that have been noted in the foregoing chapters. The latter issue is addressed in the final chapter.

The Bureaucratic Game

The government-agency bureaucrats and the scientific researchers may be caught in Hartle's (1983) "bureaucratic game." The con-

10: Scientific Validity and Political Sensitivity 135

cern here is that if research results do not confirm the prior beliefs of the project directors or the policy position of the current administration, the research may not see the "light of day." Results from in-house government research or non-government research performed using government funding may simply be suppressed. Outside research groups that depend upon government research funding are under pressure to do research and to generate results that will be received favorably by the respective funding agency. Instead of searching for "truth," researchers may be justifying prior policy positions taken by others in a position to influence their well-being. Some examples will best demonstrate these problems.

The economics chapter of the 1983 critical assessment review of acid deposition for the EPA was dropped, apparently, because the author of the chapter was critical of the biological (mainly agricultural) research that was being performed in this area. The author was critical of the relevance of the research as an input to an economic assessment and had suggestions for an improved approach (from an economic perspective). To my knowledge, the EPA report on economic impacts of acid precipitation first completed in 1980 has not been formally released by EPA as of 1993, although the contained estimates have been quoted elsewhere (Crocker and Regens 1985; Regens and Rycroft 1988).

This suppression of information in this area is not confined to the U.S. government. The socioeconomic studies on acid rain conducted for the Ontario Ministry of the Environment have never been officially published by that agency. They were made available to the public in mid-1984. These studies had been completed 2 years earlier and were made available only upon request, which means you had to know of their existence to gain access. This is hardly a wide dissemination of knowledge!

In late 1985 I was informed by a project director at Environment Canada (personal conversation) that his agency was no longer interested in economic-benefit studies of acid rain control, since the policy decision "to go ahead" had been made anyway. They obviously did not want information that might question this policy stance.

These previous examples concern the suppression of information. The acid rain phenomenon has also witnessed creative re-shaping of information or views.

The summary of the phase 2 interim report of Work Group 1 under the Memorandum of Intent (MOI 1981) contained the following conclusion from the Aquatics Group:

Atmospheric loadings of acid measured in terms of wet sulfate loadings in the range of 0–10 kg/ha/yr have been suggested as being protective for all surface waters. Wet sulfate loadings values of 11–20 kg/ha/yr have been suggested as being protective for the majority of surface waters. . . . Achieving levels suggested for moderately sensitive lakes may require a reduction in sulfate deposition of more than half.

When the final (phase 3) report of the MOI was released in 1983 (1 year late!), the following conclusion appeared:

The *Canadian* members of the Work Group propose that present deposition of sulphate in precipitation be reduced to less than 20 kg/ha/yr in order to protect all but the most sensitive aquatic ecosystems in Canada. In those areas where there is a high potential to reduce acidity and surface alkalinity is generally greater than 200 μeq/L the Canadian members recognize that a higher loading rate is acceptable (emphasis added).

This conclusion clearly suggests that the American members had withdrawn their support for the loading rates that had been proposed in the interim report. According to a spokesman for Environment Canada, the American contingent believed that there were too few data points to support the earlier conclusion. In the phase 3 report the Americans concluded that

based on this status of the scientific knowledge the U.S. Work Group concludes that it is not now possible to derive quantitative loading/effects relationships.

No explanation is offered as to why their conclusion changed between the phase 2 and phase 3 Reports. If there was consensus in phase 2, why did it change in phase 3? If new information had emerged between reports, surely this would have been used as a rationale for the divergence of opinion that appeared. No such rationale, nor even an acknowledgement of the changed position, is offered. This leads one to wonder if political sensitivity was as important (or more so) as scientific validity.

The prestigious Royal Society of Canada (1984) in its peer review of the MOI claimed that the sulphate target loading of 20 kg/ha/year, "follows naturally from the *agreed* text of the Work Group

10: Scientific Validity and Political Sensitivity

Report." The Royal Society said it could not reconcile the agreed text with the American conclusion (Melamed 1983). They further suggested that the opposing political views of the scientists' respective countries hurt the work groups (Melamed 1983). The society confessed its own bias but expressed the view that the Canadian target was too lax, noting that the Swedish target was 15 kg/ha/year and for sensitive areas even lower, 9 kg/ha/year (Melamed 1983).

It has been suggested that the work of the National Acid Precipitation Assessment Program (NAPAP) has been subjected to political re-interpretation. The NAPAP 5-year interim report to Congress was due in 1985. However, the 1985 annual report as well as the 1985 interim assessment report were not released until 1987.

Congressman John Dingell (D. Mich.) asked the Government Accounting Office (GAO) to investigate the reason for NAPAP's delays. The GAO concluded that the delays were due to staffing constraints and extensive reviews by the new director appointed in 1985 (*Coal Week* 1987b). The new director, Lawrence Kulp, oversaw a substantial revision of the interim report. One researcher is quoted as saying of Kulp: "Larry has strong opinions that may or may not be true . . ." and an administration official is quoted as saying, "Larry is pushing *his own views* on oxidants . . ." (emphasis added) (Science 1987).

Another researcher claims the NAPAP report represents an "unusual selective elimination of research" and incorporates "research of questionable value and high uncertainty" (*New York Times* 1987). The science reported in the main body of the report is thought to be of good quality; the objection is to the way the results are interpreted, particularly in the executive summary. Various researchers quoted in the media have complained that the executive summary (which is what the media and policymakers read) was written in a fashion that biases the inferences to be drawn. Dr. Schindler, an aquatic ecologist with Environment Canada, is quoted as saying, "It is the political people in NAPAP who use the numbers differently. . ." (Science 1987). Dr. Gene Likens, one of the early acid rain researchers, is quoted as saying, "It [the NAPAP report] seems rather highly politicized . . ." (Science 1987).

The NAPAP report has been criticized for using a lake pH of 5.0 as its definition of an acidified lake. As discussed in Chapter 5, aquatic damages occur at pH >5.0. In criticizing the NAPAP report, Schindler notes evidence of damage at pH 6.0, and notes that most fish species have ceased reproduction at a pH 5.3 or 5.6. According to Schindler, by the time the lake pH has dropped to pH 5.0, 30–50% of the lake's natural biota have been lost (Science 1987). The eminent acid rain ecologist, Dr. Eville Gorham, suggested that a lake pH of 5.5 would have been a better definition than the NAPAP value of 5.0. Furthermore, at pH 5.5, Gorham believed that 20% of lakes in sensitive areas would be classified as acidic (*New York Times* 1987). Kulp is quoted as defending the pH 5.0 because this is the level that sport fish begin showing dramatic effects and while other organisms may be affected at higher pH, these damages did not lower the economic value of the freshwater body (*New York Times* 1987). The research quoted in Chapter 5 shows that, contrary to Kulp's assessment, sport fish do cease reproduction at pH >5.0. Kulp offers no rationale as to why other organisms have no economic value. Indeed, under Kulp the NAPAP program pulled back from doing economic analysis of the subject (Science 1987)!

The NAPAP executive summary reveals some very subtle changes in wording that reduce the significance of some impacts. In general, the effects sections of the executive summary are taken *word-for-word* from the conclusion of the respective effects chapters. However, there are some subtle changes.

The conclusion in the chapter on aquatics and the executive summary both claim, with adequate data, that 9% of the lakes in the Adirondacks were judged to have lost several acid-sensitive species. The conclusion goes on to say, *"smaller numbers of fishless waters* have been identified in northern New England, Massachusetts and Pennsylvania" while the executive summary refers to *"a smaller percentage of fishless waters"* which need not be the same thing (emphasis added).

In assessing the direct impacts of acid rain upon human health, the conclusion in the main volume contained a brief paragraph dealing with uncertainties and noted a lack of information on "the levels of exposure to acidic aerosols for various population groups across North America; chronic health problems resulting from short-term changes in respiratory symptoms and decrements in lung function; and the effects resulting from repetitive or long-term exposures to air pollutants." The executive summary did not acknowledge these shortfalls.

When discussing the indirect effects on human health, the concluding section of the main volume states, "a direct causal relation-

10: Scientific Validity and Political Sensitivity 139

ship between rainfall acidity and levels of mercury in fish has not been established." For some reason the executive summary drops the word "direct."

The conclusion on the visibility effects begins, "The pollutants primarily responsible for reductions in visibility are *sulfate and nitrate aerosols,* carbonaceous materials. . . ." The executive summary for this section begins, "The pollutants primarily responsible for reductions in visibility are *carbonaceous materials* including soot from combustion and organic carbon (dominant in the West), sulfate and nitrate aerosols (major factors in the East) . . ." (emphasis added).

The reversal of ordering is interesting, since the executive summary places the acidic compounds much further down in what appears to be a priority listing. Furthermore, the executive summary includes in parentheses the key region for the various pollutants that the equivalent sentence in the main volume did not. The main volume did, however, include a paragraph that associates pollutants and visibility reduction in the various regions. This paragraph, which was not included in the executive summary, reads, "Sulfate particles are the *dominant* chemical components in the air responsible for non-weather related reductions in visibility in the rural East. In the West, and in urban areas of the East, carbonaceous material *may also* be a dominant factor" (emphasis added).

The main volume, unlike the executive summary, does not claim that carbonaceous material *is* a dominant factor in the West but that it *may also be* a dominant factor—in addition to sulfate particles! Furthermore, the main volume identifies sulfate particles as the *dominant components,* not sulfate *and* nitrate as *major* factors as in the executive summary (emphasis added)!

The main volume on effects has a general discussion of approaches to assessing the economic damages to materials from acid rain but provides no estimates. The executive summary in a section removed from its discussion of the materials effects states, "There may be a benefit in the reduction of acidic precipitation and sulfur dioxide concentrations on certain materials; it has not been possible to quantify this effect for this report." However, studies conducted for NAPAP on materials damages, with quantitative and economic estimates, were discussed in Chapter 6 of this book! These studies were available in the published literature. Why does the NAPAP report choose to ignore them?

The report concludes that emissions will decline or hold steady and that mandated controls may not be required, since the emerging technologies are so efficient that the utilities may voluntarily adopt them. This view, however, is in sharp contrast to that of an EPRI analysis, which concluded that *"No coal technology* offers either substantial generic capital cost or power cost reduction *compared with prevailing conditions"* (Moore 1987) (emphasis added).

The NAPAP interim report was to have included policy recommendations according to its mandate. The report did not address this issue, since it felt the uncertainties were too great to reach conclusions (*Coal Week* 1987b).

Another example of revised interpretation of scientific results is found in an unreleased 1987 EPA report on New England lake acidification. An EPA *staff* report, which concluded that 300 lakes in Connecticut, Rhode Island, and southern Massachusetts were likely to be acidified within 50 years in the absence of a reduction in SO_2 emissions, was re-analyzed by a special group of scientists and acid rain experts advising EPA. This group concluded that the EPA staff read more into the data than was appropriate (*Coal Week* 1987a). An EPA spokesman suggested in light of the re-assessment that the number of lakes in the Northeast could range from zero to a few hundred! *Coal Week* (1987a) reported that this study may never be officially published; hence, it can never be subjected to objective review!

Agency bureaucrats have an incentive to agree to further research on acid rain in order to increase their department budgets. If there are research funds available, there will always be scientists, academic or in consulting agencies, who can design some study to use the research funds. As Scott (1986) has noted, research efforts in this area "are part of an unbounded, endless, research domain" with almost any inquiry in a number of academic disciplines being justified as relevant. Furthermore, scientists on government funding may come under pressure, either directly or indirectly, to support the policy positions of the respective funding agency. One acid rain researcher stated that there are "political pressures on everyone not to say anything—people's careers are on the line" (*New York Times* 1987).

The Canadian government research agenda also has had problems. In 1984 the Canadian government asked the Royal Society of Canada to conduct a peer review of the government research program on the Long Range Transport of Air Pollutants in North America (LRTAP). This review included both federal agency in-house re-

10: Scientific Validity and Political Sensitivity 141

search and federal agency–funded external research. The society was quite critical of the program and criticized the emphasis on acid deposition and sulfur in particular. They noted that the international research thrust has been toward an integrated study of a "complete roster of pollutants including most notably oxidants such as ozone" (Royal Society of Canada 1984). They termed the relative neglect of nitrogen oxides unwise, but noted that the emphasis was "clearly a response to political priorities." The society, again in this peer review, suggested that "political controversy between the two countries has been a handicap to joint Canadian–U.S. research." They point out that "there is often a drastic mismatch between what politics needs and science can offer, at least in a hurry."

The Royal Society also criticized the Canadian government for making "too little use of university scientists in the LRTAP program," noting that in recent years federal policy favors contracting out to private consultants rather than to universities. Since the Royal Society is predominantly a prestigious *academic* group, one might attribute this to self-serving academic interests; however, the view was expressed especially by the U.S. reviewers who would not necessarily benefit from a policy change favoring the funding of university research by Canadian government agencies. It is tempting to suggest that the federal policy helps the government keep studies "under wraps" if desired. The academic researcher's career is built upon the publication and wide dissemination of research results. Academic freedom to publish is the cornerstone of the research university. The government agencies would find it harder to suppress results because it would have to violate this academic freedom. On the other hand, with a report from a private consultant, the government in essence is buying the report. The consulting firm gets its money and credit for performing the study, and is less concerned about "official publication" of the results.

This policy of preferring in-house or private research can also have an effect upon the quality of research that is produced. The academic researcher is accustomed to peer review as a natural part of the research to publication process. The research is subjected to rigorous review by anonymous referees prior to publication in academic journals. Once published it is subjected to review by the academic profession at large. This ensures that the research is judged by the highest standards of the day. In-house government research and consultant's reports may not get this rigorous inspection, and hence, they are protected and the quality of research may suffer. The Royal Society in fact said that such external reviews "are common in the universities, but we suspect, *unfamiliar to most of those who appeared before us*" (emphasis added).

The aquatics and atmospheric science programs generally received approval. The aquatics work was judged to be of very high quality; however, its future funding was in question at the time of the peer review. The terrestrial program was criticized for being nonrigorous and for being "directed towards impossibly large objectives."

The socioeconomic program was generally deemed unsatisfactory. The work was "excessively concerned with methodological review; most [governmental] departments seemed to lack a corporate memory for past studies." Further, "it ignored (*or was unaware of*) well-established policy analysis methods" (emphasis added). Stronger still, some presenters *"did not appear to fully understand* what values were being investigated or *to see clearly,* how they might be estimated" (emphasis added).

The reviewers were less than satisfied with the controls program noting that emphasis in some projects was *"political* rather than technical" (emphasis added).

Given the incentive structures faced by the agency bureaucrats and scientists, there is reason to believe that further research into the impacts of acid rain (or on clean-coal technology) will be less than satisfying as a policy-decision input. The frustration in this area is echoed by Congressman Scheuer (D. N.Y.) who states, "We're engaged in acid rain research up to our gazoo" and by Congressman Roe (D. N.J.) who states, "We've spent $400 million and we haven't done anything worth a tinker's damn" (Engelau 1987). As EPA administrator, William Ruckleshaus claimed in 1984:

> The amount of evidence necessary to convince you if you live in New England and you think Indiana is going to pay for it is a lot less than if you live in Indiana and you think New England is going to benefit from it (U.S. News & World Report 1984).

References Cited

Coal Week. 1987a. March 30.
———. 1987b. July 13.
Crocker, T.D., and J.L. Regens. 1985. Acid deposition control: A benefit-cost analysis: Its prospects and limits. Environ. Sci. Technol. 19:112–16.
Engelau, D. 1987. Washington report. J. Air Pollut. Control Assoc. 37:776.
Hartle, D.G. 1983. The theory of 'rent-seeking': Some reflections. Can. J. Econ. 16(4):539–54.
Melamed, D. 1983. Washington report. J. Air Pollut. Control Assoc. 33:812.
Memorandum of Intent on Transboundary Air Pollution (MOI). 1981. Phase 2 interim working paper of the impact assessment work group. Ottawa and Washington, D.C.
———. 1983. Final report of the impact assessment work group. Ottawa and Washington, D.C.
National Acid Precipitation Assessment Program (NAPAP). 1985. Annual Report. Washington, D.C.
———. 1987. Interim Assessment Report, 1985. Washington, D.C.
New York Times. 1987. Government acid rain report comes under attack. (September 22): 19, 22.
Regens, J.L., and R.W. Rycroft. 1988. The Acid Rain Controversy. Pittsburgh, Pa.: University of Pittsburgh.
Royal Society of Canada. 1984. Long-range transport of air pollutants in North America: A peer review of Canadian federal research, March 1984.
Science. 1987. Federal report on acid rain draws criticism. (September 18): 1404–6.
Scott, A.D. 1986. The Canadian-American problem of acid rain. Nat. Resour. J. 26:338–58.
U.S. News & World Report. 1984. After the turmoil at EPA—The successes and the failures. June 18.

CHAPTER **11**

Non-Convexities, Irreversibilities, and Acid Rain Controls as Insurance

Crocker and Forster (1981, 1986) note that the natural science literature reveals certain qualitative features that support prompt rather than delayed action on acid rain controls. The first of these is referred to as a non-convexity in the damage function, and the second is the irreversibility of some of the physical impacts, which produces a ratchet effect in the damage function.

In conventional economic analysis it is assumed that as pollution loadings rise, the additional damage becomes progressively larger with each unit increase in pollution. In the language of the mathematician, the damage function is (strictly) convex. This relationship between additional damages (called marginal damages, MD) and pollution loading is shown in Figure 11.1.

The higher the level of pollution loading is allowed to be, the lower will be the additional control costs (called marginal control costs, MCC) of reducing pollution by one unit. The lower the level of pollution loading that is desired, the higher will be the MCC of reducing pollution by one unit (Fig. 11.1).

At A_1, the MCC of reducing pollution exceed the level of MD of increasing pollution and a higher pollution level is warranted. At A_3, the reverse is true: MD of increasing pollution levels exceed the MCC

11: Non-Convexities, Irreversibilities, and Acid Rain Controls

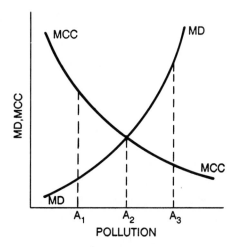

Fig. 11.1. Marginal damage and marginal control costs curves.

of a reduction in pollution and hence a lower level of pollution is desired. At A_2, the MD is equal to the MCC cost and hence A_2 is called the "optimal" pollution loading.

Crocker and Forster (1981) point out that for some of the acid rain impacts, the damage function may not produce the smooth upward sloping MD curve of textbook analysis. They cite the following tables from Butler et al. (1973).

The data in Table 11.1 suggest that in the first 0.1 decrease in pH, there is a loss of 12 stream sections with fish, while for the next 0.4 decrease in pH, there is a further loss of 6 stream sections. In the next 0.6 decrease in pH, the loss is 3 stream sections with fish. If the damage measure is "stream sections with fish lost," the additional damage is becoming smaller, not larger, as the pollution loading (acidity) is increasing.

Table 11.1. Sections with fish at various pH levels for a sample of Pennsylvania streams suffering from acid mine drainage

pH	Stream Sections with Fish
6.4	24
6.3	12
5.9	6
5.3	3
4.6	1
4.5	0

Source: Butler et al. (1973), p. 112.

Table 11.2 shows a similar result in terms of the number of species present. When the pH drops below 6.5 into the 6.4–6.5 range, there is a loss of 68 species. When the pH drops into the 6.2–6.4 range there is an additional loss (MD) of 7 species. The largest loss occurs in the early acidification stages. That is, the MD is high at low pollution loadings and decreases as the loading increases. The damage function is said to be non-convex.

In Figure 11.2 the loading level A_2 has MCC equal to MD but is not an optimal pollution level. If the loading is slightly higher, say, at A_3, the MCC is greater than MD and the pollution loading should be allowed to increase since the additional damages incurred fall short of the additional savings in control costs. If the level is A_1 then MD exceeds MCC and a lower loading level is dictated. Nonconvexities in damage functions tend to lead to prescriptions for either very stringent control responses or very lax control responses.

Figure 11.2 illustrates the MD curve for such a non-convex damage function.

If an acid rain controls program is not enacted, continued acidification could push aquatic systems beyond their equivalent of A_2. The nonconvexity of the damage function produces an "economic" irreversibility threshold at A_2. That is, it is not economically feasible to restore the environment if pollution exceeds this level.

The natural science literature also reveals a concern that some impacts may be irreversible in a physical or technical sense. The loss of genetically unique aquatic species is one example. The damage to

Table 11.2. **Variation of number of fish species with respect to pH levels for a sample of Pennsylvania streams suffering from acid mine drainage**

pH	Number of Species Present
≥ 6.5	116
$6.4 \leq pH < 6.5$	48
$6.2 \leq pH < 6.4$	41
$6.1 \leq pH < 6.2$	36
$6.0 \leq pH < 6.1$	34
$5.9 \leq pH < 6.0$	18
$5.6 \leq pH < 5.9$	12
$5.5 \leq pH < 5.6$	10
$5.2 \leq pH < 5.5$	9
$5.0 \leq pH < 5.2$	8
$4.7 \leq pH < 5.0$	7
$4.6 \leq pH < 4.7$	5
≤ 4.6	0

Source: Butler et al. (1973), pp. 96–99, 114.

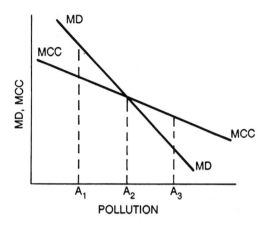

Fig. 11.2. Marginal damage and marginal control costs curves with a non-convex damage function.

historically, or culturally, significant artifacts is another. In the forest damage category there is concern that the acid deposition may damage soil quality and that these soil damages may be irreversible. Waiting for research to produce conclusive, quantitative physical or economic information concerning the nature of such damages may allow damage to be caused during the learning period that cannot be undone once the information has been generated. Crocker and Forster (1986) conclude that when planning acid deposition control strategies, "the decision maker should resist the temptation to wait until an explanation has emerged with well-defined parameters." The cautious strategy is one that will avoid such potential economic and physical irreversibilities.

Benefit-Cost Estimates and Insurance

The only published estimate of aggregate benefits of controlling acid deposition is the Crocker (1980), annual estimate of $5 billion ($ 1978). This estimate is in the range of control-cost estimates reviewed in Chapter 9. Crocker (1982) warned that there are reasons to expect benefits estimates to be biased downward and the discussions in Chapters 3 to 7 have shown the dispersion of benefit estimates for the various categories. The discussion in Chapter 9 showed that there are reasons to expect that estimates of control costs will be exaggerated or biased upward. Chapters 8 and 9 also show the possible dispersion of cost estimates.

Even if the expected benefits were to fall short of control costs, a risk-averse society may wish to implement a controls program in order to avoid potentially irreversible damages. Bearing such risk imposes a cost on society; relieving these risks is one of the benefits of control that needs to be factored into the analysis. The controls program may be seen as an "insurance premium" paid against the potential risk of irreversibility.

This argument may be clarified by using an analogy to house and automobile insurance purchased by individuals. People purchase insurance in order to protect their wealth from being diminished or destroyed by events beyond their control. It is not possible to wait until your house has burned or you have a car accident to take out the insurance. It would be too late; the personal wealth would have been lost forever.

This insurance analogy was first given by the author as part of a luncheon address to the Acid Rain in the West Conference in Gunnison, Colorado, in July 1984. During his luncheon address the following day, Charles Elkins, then an assistant administrator of EPA, challenged the "insurance" scheme by noting that there are a large number of insurance policies available that he does not take up. Nonetheless, the point remains that failure to insure one's wealth implies a gamble that it will not be lost and the saving is the insurance premium. Failure to implement an acid rain controls program implies a gamble that the damages to ecosystems, society's natural wealth, will not be irreversible.

It is true that at the end of the insured period, if the house has not burned and car accidents have been avoided, we may regret having taken out the insurance during that period, but we still renew the policy for the upcoming period to avoid the risk. In the case of acid rain controls it is also possible that in the future we will learn that damages are not as great as was feared, that the effects are not irreversible, or that the effects are not even attributable to acid rain. This is the equivalent of not having an accident during an insured time period. However, the equivalence is not complete. In the "individual" insurance example, an individual pays the insurance premium to protect his/her own wealth from adverse outcomes. In the acid rain case,

11: Non-Convexities, Irreversibilities, and Acid Rain Controls 149

the cost of the controls program may be borne by individuals or regions different from those who will reap the potential benefits of the controls program.

The electric utilities, the coal companies, the coal miners, and the electricity consumers in the upper Ohio Valley are being asked to pay for a controls program that will generate potential benefits that flow to residents in New England, New York, and Canada. Those residents who benefit from the controls program should be prepared to ease the burden of the controls program imposed upon the upper Ohio Valley.

Forster and Rees (1983) have demonstrated that it may be in the best interest of society to subsidize industries that are making substantial adjustments as a result of adverse changes in their market conditions. In the long run, economists argue that factors of production can be re-allocated to maintain full employment, while in the short to intermediate run, disruption can occur with a consequent loss in national production as well as personal hardship for those most directly affected. Subsidizing the impacted areas is designed to ease the adjustment process and reduce the loss in production.

The cost to society will be further reduced by permitting freedom of choice in abatement strategies and by phasing in the reductions over time while continuing the research efforts into the effects of acid rain. Phasing in the program will allow it to be more easily terminated if research should indicate the effects are not serious. Allowing freedom of choice encourages the development of improved technology and allows firms to adjust their abatement strategies through time. Such a phase-in controls program was advocated by and outlined in Tschinkel (1984). Scott (1986) suggested the use of pollution certificates to control acid rain precursor emissions in order to achieve the freedom of choice of abatement strategy. He also suggested that the allowable amount of emission per certificate could decline over time as a way of phasing in the program.

Much of the last two chapters has been concerned with the incentive structures facing individuals if they act in their own best interests. It is fair to end by noting that the call for acid rain controls, using a freedom-of-choice, phased in approach is being made by a risk-averse Canadian, who happens to be an economist living in Wyoming, a major low-sulfur-coal-producing state.

References Cited

Butler, R.L., E.L. Cooper, J.K. Crawford, D.C. Hales, and W.G. Kimmel. 1973. Fish and food organisms in acid mine waters of Pennsylvania. U.S. EPA Ecological Research Series EPA-R3-73-032. Washington, D.C.

Crocker, T.D. 1980. Economic impact of acid rain: Hearings before the select committee on small business environment and committee on public works. 96th Congress, 2d Session, pp. 100–11.

———. 1982. Conventional benefit-cost analysis of acid deposition control are likely to be misleading. In Acid Rain: A Transjurisdictional Problem in Search of a Solution. Ed. P.S. Gold, pp. 76–91. SUNY, Buffalo: Canadian-American Center.

Crocker, T.D., and B.A. Forster. 1981. Decision problems in the control of acid precipitation: Nonconvexities and irreversibilities. J. Air Pollut. Control. Assoc. 31:31–37.

———. 1986. Atmospheric deposition and forest decline. Water, Air, Soil Pollut. 31:1007–17.

Forster, B.A., and R. Rees. 1983. The optimal rate of decline of an inefficient industry. J. Public Econ. 22:227–42.

Scott, A.D. 1986. The Canadian-American problem of acid rain. Nat. Resour. J. 26:338–58.

Tschinkel, V. 1984. Acid rain options. J. Air Pollut. Control Assoc. 34:627–28.

Epilogue

In July 1989, in the pristine environment of Jackson, Wyoming, President George Bush unveiled his proposal for amending the Clean Air Act. The president's proposal contained a section devoted to reducing the precursors of acid rain. Sulfur dioxide (SO_2) emissions in particular were to be reduced by 10-million tons from 1980 base levels. The proposal featured a phase-in approach with emissions trading and emissions fees included.

With the president coming forward with his own acid rain control proposal, the political landscape on this issue was transformed 180 degrees. Rather than blocking independent bills as it had for close to a decade, the Congress moved to bring forward bills of its own. In less than a year from the president's Jackson proclamation, both the Senate and House had voted to approve clean air bills that contained acid rain control provisions. The strength of the voting shows just how dramatic the shift had been. The Senate bill (S. 1630) passed on April 4, 1990, by a vote of 89–11. The House bill (H.R. 3030) passed on May 23, 1990, by an overwhelming vote of 401–21. Within 5 months of the House approving its bill, the joint congressional negotiators had reached agreement on a compromise bill to be forwarded to the president for his approval.

While the Senate and House bills were being debated and during the negotiation of a compromise bill, the "games" continued. The Edison Electric Institute estimated that the House bill would reduce SO_2 emissions by 13-million tons—30% more than the president's proposed target. Estimates of the House bill by EPA and the National Acid Precipitation Assessment Program (NAPAP) suggest lower reductions of 10.3-million tons and 11.3-million tons respectively.

In the Senate, Senator Robert Byrd (D. W.Va.) attempted to

The discussion in this epilogue is based upon coverage in various issues of *Coal Outlook*.

amend S. 1630 to include a coal miners' compensation package. Byrd originally proposed a $700-million package but lowered the cost to $500 million. This amendment was very narrowly defeated on a 50–49 vote. A similar amendment to H.R. 3030, proposed by another West Virginian, Congressman Bob Wise (D.), was successful. The Wise amendment provided for an extension of 6–12 months in unemployment compensation and up to 2 years of job training. The amendment provided for a maximum of $50 million annually for 5 years. Opponents of the amendment contended that the costs could rise to $3.9 billion rather than the $250 million claimed by Congressman Wise.

Senators Arlen Specter (R. Pa.) and Howard Metzenbaum (D. Ohio) proposed an amendment to give the utilities a 20% tax credit if they opted for scrubbers. This measure was defeated.

The president had indicated that he would veto any bill that would exceed the cost of his proposal by more than 10%. The cost of the president's own proposal was initially figured at $19 billion, but this was later revised to $22 billion annually (for the full Clean Air Act amendments, including the acid rain component). The Clean Air Working Group (CAWG), an industry lobby group, argued that the least-cost package being considered by Congress would cost at least $51 billion annually, while the most expensive version would be in the $91-billion range annually. Each of these estimates is far in excess of the president's target.

The chairman of the Council of Economic Advisors, Michael Boskin, estimated the cost of House and Senate bills to be $25 billion annually and may be $35 billion if the more costly aspects were adopted. While Boskin's estimates are considerably lower than CAWG's, they lead him to believe that the president should veto such a bill because it posed a threat to the nation's economy.

CAWG estimated that the acid rain component would cost between $4 billion and $7 billion annually. NAPAP on the other hand suggested an upper bound of $4 billion with $7 billion being relevant for a larger SO_2 reduction (12-million tons rather than 10-million tons). EPA estimates an upper bound of $5 billion annually in the year 2005, with earlier costs falling below $4 billion annually.

The National Coal Association (NCA) determined that the congressional bills would result in a loss of 35,000 coal mining jobs. This contrasts with an EPA estimate of job losses of no more than 16,000 in the high-sulfur coal areas—less than half the number projected by

Epilogue

the NCA. EPA further estimates that jobs in low-sulfur coal areas will expand by at least 13,000! The NCA omits reference to job increases.

Despite all these games, the Congressional Conference Committee was able to agree on a package that the president signed on November 15, 1990. The 10-million-ton reduction in SO_2 would take place in two phases over a 10-year period. Under the phase 1 requirements, power plants must reduce SO_2 emissions to 2.5 pounds per million BTUs by January 1, 1995. The emissions must drop to 1.2 pounds per million BTUs by January 1, 2000 under phase 2. Utilities that reduce their emissions by more than required by the CAA will receive emissions credits that may be used to offset emissions at other units, or at future plants, or may be sold in an organized trading market. Plants that reduce emissions early may earn "bonus credits" on a two for one basis. The bill signed by the president also contained a version of Congressman Wise's displaced coal miner compensation package.

With the president's signing of the compromise bill, the acid rain game shifted to the utilities deciding what strategy they should adopt. The coal companies and the United Mine Workers (UMW) are attempting to influence the utilities' decisions in order to protect their own income-earning position (Table E.1.).

The market shares increase noticeably for southern Appalachia and the Rocky Mountain region. The largest market share losses occur for the Illinois Basin and northern Appalachia, in that order.

Analysis by Shearson Lehman Brothers (SLB) suggests that the low-sulfur producer, Ashland Coal, in central Appalachia will experience a 20% growth in company earnings as a result of the 1990 CAA amendments. SLB estimate a shift of 100-million tons annually from high- to low-sulfur coal.

The UMW began lobbying state legislatures to obtain incentives

Table E.1. Coal production shares by region before and after passage of bills

	Baseline (before)	House/Senate Bills (after)
Northern Appalachia	20.97%	17.75%
Southern Appalachia	25.80%	32.64%
Illinois Basin	22.05%	15.46%
Great Plains	5.44%	4.19%
Rocky Mountains	25.74%	29.96%

Source: *Coal Outlook,* September 10, 1990.

for utilities to install scrubbers rather than engage in fuel switching.

SLB suggests that railways stand to gain from the 1990 amendments. In March 1991, Burlington Northern announced plans for capital improvements to handle an anticipated increase of 50- to 60-million tons of coal coming from the low-sulfur coal mines of the Powder River Basin (PRB) in the Rocky Mountain region. Half of this anticipated increase is attributed to the 1990 CAA amendments with the other half being attributed to normal growth in coal demand. In July 1990, a coal and utility lobby group, Consumers United for Rail Equity (CURE), expressed concern that the railroads would use the CAA as an excuse to charge excessive rail rates on low-sulfur coal shipments from captive mines.

The electric utilities identified by SLB as having the most exposure to acid rain controls impacts are American Electric Power (AEP), General Public Utilities (GPU), and Allegheny Power. By early December 1990, AEP announced plans to select scrubbing as its compliance strategy. AEP was hoping to "overcomply" on some units to generate bonus emission credits, which could be used to offset emissions at some other units or to generate revenue in the new emissions-trading market. For this strategy to be worthwhile, AEP believed that an emissions credit price of $1000 per credit (a ton of SO_2) would be necessary.

By late January, early February 1991, AEP was having second thoughts and the possibility of shifting to PRB coal was coming to the forefront. According to AEP, switching to PRB coal would have a capital cost of $125/kW while capital costs for scrubbing amount to $230/kW. Switching to PRB comes in with operating costs of $115/kW annually compared to $170/kW annually for scrubbing. AEP by this time was less certain about receiving overcompliance credits than it had been in December 1990.

The UMW disagreed with AEP's cost estimates and argued that the company would be better off scrubbing. The UMW, using an emissions credit price of $1000, suggested that the credits earned would cover the bulk of the scrubbing cost. AEP countered that the UMW's estimates were unduly rosy. AEP at this point is forecasting an emissions credit price of $400 in phase 1 and $700 in phase 2. Furthermore, they argue that they might not get any bonus credits.

In December 1990, GPU suggested that it would scrub at one plant and use coal-switching at another. It was hoping to earn overcompliance credits to offset emissions at other plants. GPU also

Epilogue   **155**

wanted to land a scrubber contract quickly in order to avoid an anticipated price increase for scrubbers. By early February 1991, GPU revised its compliance-cost estimate to $675 million (down from $1 billion) because it would need fewer scrubbers than originally believed. GPU confessed to being on a "learning curve."

Allegheny Power will scrub one of its plants hoping to obtain "early scrub" bonus credits to offset other plants. Allegheny believes that scrubbing is its least-cost strategy. According to its own estimates, the overall costs involved in coal-switching amount to $840 per ton of SO_2 removed, while scrubbing leads to a cost of $700 per ton of SO_2 removed. Allegheny has a scrubber contract with General Electric with a cost of $725 million.

Northern Indiana Public Service Company is likely to use coal blending with low-sulfur Wyoming coal. In the short term, Pennsylvania Power and Light will use coal-cleaning and coal-switching to meet the targets. In phase 2 they may need to opt for scrubbers. Indianapolis Power and Light will scrub and overcomply to obtain credits to offset emissions at some plants. Illinois Power prefers scrubbers and using in-state coal. It hopes to cover other plants by over-complying.

What is clear in observing the behavior since the president signed the Clean Air Act amendments in November 1990 is that the acid rain debate continues. The rules of the game have changed and the alliances have been altered, but self-interest will continue to influence the choice of control strategy. The flexibility of choice afforded by the 1990 CAA will enable society to achieve the reduction in SO_2 emissions in a less costly manner than if the legislation had followed the pattern of the 1977 amendments. For this, a vote of thanks to the Bush administration and the Congress from the present and future generations is due.

Selected General References

Adams, R.M. 1986. Agriculture, forestry and related benefits of air pollution control: A review and some observations. Am. J. Agric. Econ. (May): 464–72.
Adams, R.M., J.M. Callaway, and B.A. McCarl. 1986. Pollution, agriculture and social welfare: The case of acid deposition. Can. J. Agric. Econ. 34: 3–19.
Altshuller, A.P., ed. 1984. The Acidic Deposition Phenomenon and Its Effects: Critical Assessment Review Papers, Vol. 1, Atmospheric Sciences. EPA-600/8-83-016AF, U.S. EPA, Washington, D.C.
Baker, J.P., and T.A. Haines. 1986. Evidence of fish population responses to acidification in the eastern United States. Water, Air, Soil Pollut. 31:605–31.
Brocksen, R.W., and A.S. Lefohn. 1984. Acid rain effects research—a status report. J. Air Pollut. Control Assoc. 84:1005–13.
Callaway, J.M., R.F. Darwin, and R.J. Nesse. 1986. Economic valuation of acidic deposition damages: Preliminary estimates from the 1985 NAPAP assessment. Water, Air, Soil Pollut. 31:1019–34.
Cowling, E.B. 1984. What is happening to Germany's forests? Environ. Forum (May): 6–11.
Crocker, T.D. 1985. Acid deposition control benefits as problematic. J. Energy Law Policy, 339–56.
Crocker, T.D., J.T. Tschirhart, R.M. Adams, and B.A. Forster. 1980. Methods development for assessing acid precipitation control benefits, a report to the U.S. EPA.
Crocker, T.D., and B.A. Forster. 1981. Decision problems in the control of acid precipitation: nonconvexities and irreversibilities. J. Air Pollut. Control Assoc. 31:31–37.
Evans, L.S., G.R. Hendry, G.J. Stensland, D.W. Johnson, and A.J. Francis. 1981. Acidic precipitation: Considerations for an air quality standard. Water, Air, Soil Pollut. 16:469–509.
Fay, J.A., D. Golomb and S. Kumar. 1985. Source apportionment of wet sulfate deposition in eastern North America, Atmos. Environ., 1773–82.
Forster, B.A. 1984. An economic assessment of the significance of long-range transported air pollutants for agriculture in eastern Canada. Can. J. Agric. Econ. 32:498–525.

———. 1985. Economic impact of acid deposition in the Canadian aquatic sector. In Acid Deposition. Ed. D.F. Adams and W.W. Page, pp. 409–37. New York: Plenum Press.

———. 1989. Acid rain games: Incentives to exaggerate control costs and economic disruption. J. Environ. Manage. 28:349–60.

Forster, B.A., and T.P. Phillips. 1987. Economic impact of acid rain on forest, aquatic, and agricultural ecosystems in Canada. Am. J. Agric. Econ., 963–69.

Freda, J. 1986. The influence of acidic pond water on amphibians: A review. Water, Air, Soil Pollut. 30:439–50.

Gallogly, M. 1981–1982. Acid precipitation: Can the Clean Air Act handle it? Boston Coll. Environ. Aff. Law Rev. 9(3):687–744.

Garland, C. 1988. Acid rain over the United States and Canada: The D.C. circuit fails to provide shelter under Section 115 of the Clean Air Act while state action provides a temporary umbrella. Boston Coll. Environ. Aff. Law Rev. 16:1–37.

Glauthier, T.J., and F. Mayer. 1987. A critical perspective on NAPAP's methodology for materials damage assessment. J. Air Pollut. Control. Assoc. 37:683–84.

Horst, R.L., T.J. Lareau, and F.W. Lipfert. 1987. Economic evaluation of materials damage associated with acid deposition. J. Air Pollut. Control Assoc. 37:682–83.

Irving, P.M. 1983. Acidic precipitation effects on crops: A review and analysis of research. J. Environ. Qual. 12:442–53.

Jones, D.N., and R.A. Tybout. 1986. Environmental regulation and electric utility regulation: Compatibility and conflict. Boston Coll. Environ. Aff. Law Rev., 14:31–59.

Krupa, S.V., and A.S. Lefohn. 1988. Acidic precipitation: A technical amplification of NAPAP's findings. J. Air Pollut. Control Assoc. 38:766–76.

Lewis, D., and W. Davis. 1986. Joint report of the special envoys on acid rain.

Linthurst, R.E., ed. 1984. The Acidic Deposition Phenomenon and Its Effects: Critical Assessment Review Papers, Vol. 2, Effects Sciences. EPA-600/8-83-016B, U.S. EPA, Washington, D.C.

McDonald, M.E. 1985. Acid deposition and drinking water. Environ. Sci. Technol. 19:772–76.

McLaughlin, S.B. 1985. Effects of air pollution on forests: A critical review. J. Air Pollut. Control Assoc. 35:512–34.

Maulbetsch, J.S., M.W. McElroy, and D. Eskinazi. 1986. Retrofit NO_x control options for coal-fired electric utility power plants. J. Air Pollut. Control Assoc. 36:1294–98.

Menz, F.C., and J.K. Mullen. 1985. The effects of acidification damages on the economic value of the Adirondack fishery to New York anglers. Am. J. Agric. Econ., 112–19.

Office of Technology Assessment (OTA). 1984. Acid rain and transported air pollutants: Implications for public policy. U.S. Congress. OTA-0-204, Washington, D.C.

Selected References

Pierce, B.A. 1985. Acid tolerance in amphibians, Bioscience 35:129–243.
Pierson, W.R., and T.Y. Chang. 1986. Acid rain in western Europe and northeastern U.S. — a technical appraisal. CRC Critical Rev. Environ. Control 16:167–92.
Regens, J.L., and R.W. Rycroft. 1986. Options for financing acid rain controls. Nat. Resour. J. 26:519–49.
Research Consultation Group. 1979. The LRTAP problem in North America: A preliminary overview. U.S. Dep. State and Can. Dep. Ext. Affairs.
_____. 1980. Second report on the long range transport of air pollutants, U.S. Dep. of State and Can. Dep. Ext. Affairs.
Rosseland, B.O. 1986. Ecological effects of acidification on tertiary consumers. Fish population responses. Water, Air, Soil Pollut. 30:451–60.
Scott, A.D. 1986. The Canadian-American problem of acid rain. Nat. Resour. J. 26:338–58.
Shepard, M. 1988. Coal technologies for a new age. EPRI J. (January/February): 4–17.
United States–Canada Memorandum of Intent on Transboundary Air Pollution, various work group reports.
Willett, K. 1986. Environmental management costs using a best available control technology BACT in the electric generating industry. Manage. Decis. Econ. 7:29–36.

Index

Abatement strategies, 149
Acid deposition, 12
Acid lakes, 59, 137–138
Acid precipitation, 12. *See also* Acid rain
Acid rain
  agricultural impacts
    economic, 39–44
    physical, 30–38
  American-Canadian research efforts on, 3–4
  aquatic impacts
    economic, 64–75
    physical, 57–63
  benefit-cost analysis, 147–149
  definition of, 12, 13
  documented effects of, ix
  economic-benefit studies of, 135
  under Executive Order 12, 291: 7, 8, 117
  first description of, ix
  forest impacts
    economic, 54–55
    physical, 48–54
  health impacts
    economic, 95–96
    physical, 90–94
  indicators of, 26–27
  linearity hypothesis of, 25–26
  materials damage
    economic, 84–87
    physical, 78–84, 87
  need for prompt action, 144–147
  pH in defining, 14
  politicization of, as issue, 129–131
  precursors, 13, 101
  special interest behavior and policy formation, 8–10
  time trends in, 27
  transport of airborne pollutants, 20–25
  U.S. policy on, ix, 4–6
  visibility impacts
    economic, 95–98
    physical, 94
Acid Rain Bill H.R. 2497, 107

Acid Rain in the West Conference, 148
Administrative Procedures Act, 6
Advanced Statistical Trajectory Regional Air Pollution Control (ASTRAP) Model for sulfur deposition, 23–24
Aesthetic damage, estimating, 96–97
Agricultural soils, effects of acid rain on, 37–38, 41, 42–43
Agriculture
  economic impacts
    direct, 39–40
    indirect, 41–43
    of ozone, 43–44
  physical impacts, 30–31
    acid deposition, 31–36
    of ozone, 36–37
Ahmad, Sharon, 4
Air and Rain (Smith), ix
Airborne pollutants, transport of, 20–25
Air-flow patterns, in materials damage, 79
Air pollution
  damage to materials, 80–81
  in forest declines, 52–54
Air Pollution Control Association Conference, 87
Alkalinity-neutralizing capacity (ANC), 58
Allegheny Power, and acid rain legislation, 154, 155
Alliance for Clean Energy (ACE), 124
Alternate fuels, in electric power generation, 101
Aluminum
  effect of, on drinking water, 92
  sensitivity to acid deposition, 78, 82
  toxic effects of, on fish, 62, 67
  toxicity hypothesis, in forest damage, 51
Ambient-based permit system (APS), 109–110
Ambient discharge permit (ADP) system, 111, 112
American Electric Power (AEP), 124, 154
American Gas Association (AGA), 122
American Public Power Association, 122

161

American Southwest, value of preserving visibility in, 98
Amphibians, impact of acidification on, 63
Aquatic birds, impact of acidification on, 63
Aquatics
  economic impacts
    liming costs, 64–67
    losses, 69–74
    sportfishing losses, 67–69
  physical impacts, 57, 137–138
    aquatic sector at risk, 57–60
    on fish, 60–62
    non-fish, 62–63
ARA Consultants study, on aquatic impacts, 70–74
Asbestos, effect of, on drinking water, 92
Ashland Coal, 153
Aspin, Les, 107
Asthma, 90
Atmospheric deposition, compounds in, 12–13
Automotive exhaust standards, 17

Benefit-cost estimates, 7, 147–149
Best available control technology (BACT), 106, 117, 118–119, 124
Bidding-game approach, to estimating aesthetic damage, 96–97
Bilateral Research Consultation Group on the Long Range Transport of Air Pollutants, 3
Biological agents, in physical coal cleaning, 100–101
Birds, impact of acidification on aquatic, 63
Boskin, Michael, 152
Bronchitis, 90
Bronzing, 37
Brown leaf, 37
Bubble concept, 110
Bureaucracy, in acid rain control, 134–142
Bush administration, on acid rain regulation, ix, 151–155
Byrd, Robert, 128, 129–131, 151–152

Cadmium, effect of, on drinking water, 92
Calcite saturation index (CSI), 58–59

Calcium hydroxide, as sorbent, 102
Canada
  agricultural damage in, 40, 42, 43–44
  automobile exhaust standards in, 17
  control of SO_2 emissions, 4
  electricity threat to U.S. industry from, 130
  environmental quality in, and willingness to pay for, 73–74
  research efforts on acid rain in, 3–4, 135, 140–141
  SO_2 and NO_x emissions in, 15–18
Canadian Clean Air Act (1980), 5
Carbonates, sensitivity to acid deposition, 78, 82
Carter administration, on acid rain regulation, ix, 5
Cheney, Richard, 131
China Syndrome, 101
Chlorine, in testing for materials damage, 79
Citizens for Sensible Control of Acid Rain (CS-CAR), 9, 128
Clean Air Act (CAA) (1970), 116, 118
  on acid rain, ix
  implementation of, 6
  1977 amendments to, 105, 106, 116, 118, 123, 127
  1990 amendments to, 121, 151–155
  state implementation plans under, 5, 111, 112
Clean Air Coalition, 131
Clean Air Working Group (CAWG), 152
Coal, removal of sulfur from, 100–101
Coal miners, and control costs and disruption, 126–128
Coal suppliers, and control costs and disruption, 123–126
Command-and-control policies, for emissions reductions, 105–106
Compensation, 98
Congress, U.S., acid rain control debates in, 129–131, 137, 142, 151–152
Construction work in progress (CWIP), 121
Consumers United for Rail Equity (CURE), 154
Control costs and disruption, 116–131
Control options and strategies
  command-and-control policies, 105–106
  comparison of policies, 111–113

Index

for emissions reduction from the electric-power industry, 100–105
for implementing emissions reductions, 105
market-incentive policies, 106–111
Copper
contamination of drinking water by, 92, 95
sensitivity to acid deposition, 78, 82
toxic effects on fish, 62
Copper smelters, as source of acid rain, 23
Corn, effects of acid deposition on, 35, 40, 43, 44
Corrosion, 82
Costle, Douglas, 5, 6

Davis, William, 4
Desulfurization technologies, 102
Dialysis dementia, 92
Dingell, John, 129, 137
Dowd, Joseph, 118
Downey, Thomas, 107
Downstream injection, 102
Drought, and forest damage, 52
Dry sorbent emission control technologies, 102
Durenberger, David F., 130

Eckart, Dennis Edward, 129
Economic impacts, 8–9
in agriculture, 39–44
on aquatic life, 64–67
in forests, 54–55
on materials damage, 84–87
Edison Electric Institute (EEI), 119, 151
Electricity consumers, 128–129
Electric-power industry
as acid rain source, 16, 17, 18
control costs and disruption of, 118–122
and emissions reduction, 100–105
Electric Power Research Institute, 21, 98, 104, 124, 126
Elkins, Charles, 148
Emission discharge permit (EDP), 111, 112
Emission permit scheme (EPS), 108, 109
Emission reduction credits (ERCs), 110

Emissions reductions, 100–105, 106–111
Emissions Trading Policy Statement, 110
Emphysema, 90
Environmental-quality ladder, 71
Environmental Protection Agency (EPA)
bubble concept of, 110
and impacts of acid rain, 135
on lake acidification, 140
and State Implementation Plans, 5
and WEPCO provision, 120–121
Environment Canada, 135, 136, 137
Executive Order 12, 291, 7, 8, 117, 127

Falconbridge Mines, and SO_2, 18, 23, 24
Fish
acidification impact on, 59, 60–62, 67–69, 138, 145–146
mercury accumulation in, 92, 93–94, 95, 139
toxic metals in, 91
Flue gas desulfurization (FGD), 102, 103, 122, 123
Fluidized-bed combustion (FBC), 104
Forests
economic impacts on, 54–55
physical impacts on, 48–50
competing hypotheses of forest damage, 50–54

Gaseous pollutants, in forest damage, 50–51
Gasification combined cycle (GCC), 104
General Public Utilities (GPU), and acid rain legislation, 154–155
Germany, forest damage and decline in, 48–51
Gorham, Eville, 138
Gorsuch, Anne, 5, 22
Grand Canyon, preserving visibility in, 98
Great Britain, and SO_2 emissions, 4
Greenhouse-effect precursors, 101
Gregg, Judd, 107
Groundwater, effect of acid deposition on, 92–93

Health impacts, 90, 138–139
direct, 90–91
economic, 94–95

Health impacts (*continued*)
  indirect, 91–94
Heavy metals, and forest damage, 51

Illinois Power, and acid rain regulation, 155
Improved pulverized coal (IPC), 104
Indianapolis Power and Light, and acid rain regulation, 155
Industrial Gas Cleaning Institute (IGCI), 119–120
Insects, and forest damage, 52
Insurance analogy, in acid rain control, 148–149
International Nickel Company (INCO), as source of SO_2, 18, 21, 23, 24
International Nitrogen Oxides Protocol, 17, 106
Irreversibility, of acidification, 144, 146–147, 148
Izaak Walton League, 6

Johnson, Norma Holloway, 5

Kulp, Lawrence, 137, 138

La Cloche Mountain lakes, population losses in, 57
Lakes
  acidification of, 59–60, 137–138
  neutralization of, 64–67, 102
Lead, effect of, on drinking water, 92
Lewis, Drew, 4
Likens, Gene, 137
Lime
  applications of, in soil, 38, 41, 42
  in lake neutralization, 64–67, 102
  as sorbent, 102
Linearity hypothesis, 25–26
Lobby, against acid rain controls, 9, 128, 152
Long-range transport (LRT) models, 23
Long Range Transport of Air Pollutants in North America (LRTAP), 20–23, 25–26
  and agricultural impacts, 30
  peer review of, 140–141
Luken, Charles, 129

MacEachen, Allan, 3
Magnesium deficiencies, and acid deposition, 51
Maple syrup production, and acid rain, 49–50, 51, 54–55
Marginal control costs (MCC) of pollution reduction, 144–146
Marginal damages, 144, 145, 147
Marketable pollution right (MPR) scheme, 111
Market-incentive policies, in emissions reduction, 106–111
  versus command-and-control policies, 111–113
Materials damages, 80–81, 139–140
  economic impacts, 84–87
  physical impacts, 78–81, 82–84
McConnell, Addison Mitchell, Jr., 129
Memorandum of Intent on Transboundary Air Pollution (1980), 3, 23, 58, 78, 83, 135–136
Mercury
  accumulation in fish, 92, 93–94, 95
  effect of, on drinking water, 92
Metals, acid rain damage to, 82
Metzenbaum, Howard, 152
Mitchell, George J., 130, 131
Moore, Arch, 130
Morphoedaphic index (MEI), 61
Morris, Bob, 38
Mulroney, Brian, 3–4
Muskie, Edmund, 5

National Acidic Precipitation Assessment Program (NAPAP)
  on agricultural impacts, 34–36, 39–40, 42, 43
  on aquatic damage, 59, 67, 68–69, 126
  on forest impacts, 54
  on health effects, 91
  materials damage, assessment of, 78, 82–87
  report, 137–139
  research, 137–140
  on SO_2 emissions, 151
  soybean experiments, 34–36, 39–40
  on visibility impacts, 94
National Association of Manufacturers (NAM), and acid rain legislation, 128
National Coal Association (NCA), 123

Index

on costs of acid rain regulation, 152–153
on electricity import tax, 130
on SANE tax, 128
National Crop Loss Assessment Network (NCLAN), on ozone and crop losses, 36
National Primary Drinking Water Regulations, 92
New Source Performance Standards (NSPS), 105–106, 117
Nitrogen
  acid deposition, 51–52
  fertilization and crop yield, 38
  and forest growth, 48
Nitrogen dioxide (NO_2), 13, 90–91, 94
NO_x
  American and Canadian contributions to, 15–16
  emissions, and acid rain, 14–15
  geographical sources of, 18–19
  industrial control of, 103, 104, 105, 122
  industrial sources of, 16–18
  in materials damage, 78
  protocol, 106
  SANE tax proposal on, 107–108
Non-convexity, in the damage function, 144
Non-ferrous smelting industry, and acid rain, 16, 18
Noranda Mines, and SO_2, 18
Northern Indiana Public Service Company, and acid rain, 155
Nuclear-waste disposal, 101

Office of Management and Budget, regulatory impact analyses of, 7
Ohio River Basin Energy Study (ORBES), 43
Oxides of nitrogen. See NO_x
Ozone, 12–13
  agricultural impacts of, 31–37, 43–44
  forest damage from, 50–51, 53
  health effects of, 90

Paint/coating systems, and acid rain damage, 82–83, 86–87
pH, 13, 57, 58, 59
Pitting, 82

Poland, and control of SO_2, 4
Political representatives, regional, 129–131
Pollutants, 12, 13
Pollution
  certification, 109, 149
  loading, and marginal damages, 144, 145, 147
  offset scheme (POS), 109
  transboundary, U.S. recognition of, 4–6
Poundstone, W. N., 27
Precipitation pH, 58
Pulmonary edema, 90

Radish, acid deposition effects on, 33
Rahall, Nick J., 129–130
Rate-of-return regulation, 121, 122
Reagan administration, and acid rain policy, ix, 3–4, 5, 7, 117
Recreation, impact of acidification on, 69–74
Regulatory impact analyses, 7
Research Consultation Group, 3, 14
Respiratory ailments, 90–91
Retrofitting with emission-control technology, 102, 126
Roe, Robert A., 142
Royal Society of Canada, 136–137, 140–141
Ruckelshaus, William, 5, 129, 142

SANE tax, 107–108, 119, 128
Scheuer, James Haas, 142
Schindler, Dr., 137–138
Scientific validity and political sensitivity, bureaucratic game, 134–142
Scrubbers, 102, 110, 117
  bias toward, 121–122
  costs of, 119–120, 121, 122
Selective catalytic reduction (SCR), 103
"Shamrock" summit, 3–4
Shultz, George, 3
Sierra Club, 6
Sikorski, Gerry, 130
Silicates, sensitivity to acid corrosion, 82
Simpson, Alan K., 131
Smith, Robert Angus, ix
SO_2. See Sulfur dioxide (SO_2)
Sodium compounds, in emissions control, 102–103

Soil
  acidification of, 37–38
  acid rain effects on, 37–38, 41, 42–43
Solution acidity, pH value of, 13
Soybeans, acid deposition on, 33–35, 40, 43, 44
Special interests, in acid rain control, 129–131
Specter, Arlen, 152
Sportfishing, impacts of acidification on, 67–69
Spray drying, 102
Stafford, Robert Theodore, 130
Staggers Act, 123
State Implementation Plan (SIP), 5, 111, 112
Steel, sensitivity to acid deposition, 82
Stockman, David, 7
Stone, acid rain damage of, 82
Stress, in forest damage, 52
Stress-assisted corrosion, 82
Sulfur dioxide (SO_2)
  and acid rain, 14
  American and Canadian contributions to, 15–16
  emission permits, 113
  emissions control program, 4
  fertilization, crop yield responses to, 38
  geographical sources of, 18–19
  health effects of, 91
  industrial controls of, 104, 105, 122
  industrial sources of, 16–18
  in materials damage, 78, 82, 83
  percentage reduction of, 126–127
  as respiratory irritant, 91
  SANE tax proposal on, 107
  visibility impacts of, 94
Sulfuric acid, health effects of, 91
Sulfur Institute, 38
Surface Mine Control and Reclamation Act (1977), 123–124

Tax concessions, 121–122. *See also* SANE tax
Tennessee Valley Authority (TVA), and SO_2, 19
"30 Club," 4
Thomas, Lee, 5
Time of wetness, and materials damage, 79

Toxic metals, mobilization of, 91–92
Transferable discharge permit (TDP) system, 112
Transmission towers, effects of acid deposition on, 87
Transportation sector, as acid rain source, 16–17

Udall, Morris, 131
Union des Producteurs Agricoles, 54
United Mine Workers (UMW), 126–128, 130, 153–154
United States
  acid rain policy in, ix
  agricultural damage in, 35, 36, 39
  automotive exhaust standards in, 17
  research efforts on acid rain, 3–4, 135, 137–140
  SO_2 and NO_x emissions in, 15–18
  transboundary pollution, official recognition of, 4–5, 6
United States–Canada Research Consultation Group (RCG), 78
Utility scrub-switch decisions, 124, 125

Visibility
  economic valuation of, 95–98
  impacts, 94, 139
  reduction of, 95–96

Water, contamination of, 58–59, 91–93, 95
Weatherfleck, 37
WEPCO Provision, 120–121
Wheat, effects of acid deposition on, 43
Willingness to pay (WTP), for environmental quality, 72–74, 96–97
Wisconsin Electric Power Company (WEPCO), 120–121
Wise, Bob, 152
Wood, sensitivity to acid correlation, 78
Work Group 1 on Impact Assessment, Report of, 3

Zinc corrosion, 82
Zooplankton biomass, impact of acidification on, 63